Getting to Know Your Child's Brain

Tressa Reisetter, Ph.D.

Wisdom
Editions
Minneapolis

Wisdom
Editions
Minneapolis

SECOND EDITION DECEMBER 2022
Getting to Know Your Child's Brain
Copyright © 2017 by Tressa Reisetter. All rights reserved.

10 9 8 7 6 5 4 3 2

Cover and interior design: Gary Lindberg

ISBN: 978-1-960250-53-7

Getting to Know Your Child's Brain

Tressa Reisetter, Ph.D.

To my husband, Steve, who supported me through quitting my teaching job, getting my Ph.D., and establishing a new career in middle-age. He has been my rock for over 40 years, and I hope for many more.

Table of Contents

Introduction

Most books that work to explain technical subjects use complicated vocabulary words and are often lengthy. Let's face it. Technical subjects have specialized vocabulary, and it takes a lot of time to thoroughly explain the concepts.

On the other hand, my father was an intelligent man, but he always told me, "Never use four syllables when you can say it in one." He also told me that I had a gift for math and science, and I had a responsibility to develop and share it with mankind.

That is what I'm attempting to do with this book. I want to write something that parents who don't have PhDs can easily understand.

Over the years of working as a neuropsychologist, I have had many people thank me for explaining results to them so clearly. I am hoping that this book expands the explanations to help parents who have not come to me for a neuropsychological evaluation.

I hope to include information that will help them understand how their children's brain development and structure may be impacting their behavior, and then recommend some things they can do about it.

1—Some Basics

Nature vs. Nurture... It's Both

For decades, scientists argued about which had more effect on the development of children: their genetic code or the environment. Most experts currently agree that it is both. An individual's genetic code lays the blueprint for how that person can develop, but it is not an unavoidable destiny. What most people don't realize, however, is that interaction with the environment has a physiological effect on the brain's development.

The development of the brain is not linear. That is, it does not develop in a step-by-step, first-this-then-that kind of a way. At various stages, the brain undergoes an overabundant growth of connections, then proceeds to kill off the ones that it doesn't need in the environment it has. It's called necrosis. A two-

3

year-old has twice as many brain connections as an adult. Not half.

It's rather like the practice of bonsai. There is too much growth, then as individuals interact with their environments, the brain connections are trimmed down to what will become that individual's unique brain. For example, a child in a threatening environment will keep all the connections he needs to be safe and hypervigilant. Put that child in a school setting, and he will likely find it quite difficult to focus on what the teacher is saying, but he will know who is walking in the hall, who is whispering to his neighbor, and what the weather is like outside.

Three surges of growth occur roughly at the stages of development described by psychologist Jean Piaget in 1936. Around the age of two, one surge associated with the preoperational phase occurs. Another occurs approximately at age seven when the concrete operational stage begins. The last one occurs in the early teen years, as abstract thinking begins to develop. It is interesting that the age of two is described as "terrible," and young teens are known to be adult-like one day and childlike the next. For some unknown reason, the age of seven is not generally associated with moodiness and changes in behavior from day to day, but age two and the teen years are.

The processes of growth and killing of excess

brain connections is tough on the brain and the individual it inhabits. Contrary to what parents sometimes suspect, their children are not usually misbehaving on purpose. In fact, I believe that most children would behave well if they could. They have some control over their brains, but the control is not total. The parents ask, "What were you thinking?"

The answer is often that the child wasn't thinking at all, at least about the incident that happened. They often weren't paying attention, and that's when bad things can happen. What we can learn from these occurrences of overgrowth and paring of connections is that it is the interaction between individuals and their environment that works with the genetic blueprint to decide which connections to kill. Parents, and to a lesser degree, teachers, have a significant influence on the physiological structure of the brains of children.

Incidentally, the frontal lobe, which I call the Inner-Adult and which helps humans stay focused and make good decisions, does not mature until around age twenty for girls and age twenty-five for boys. Brain research has discovered this in the last ten years or so. However, for decades the rate for car insurance for males has decreased dramatically at the age of twenty-five. Accident rate statistics describe what science supports as the age that coincides with the maturation of the frontal lobe.

Developing Connections

While humans have around one hundred billion brain cells and potentially trillions of connections between cells, the number of connections is not set. Research has repeatedly shown that children in stimulating environments (with books, toys, opportunities to use the imagination, etc.) develop brains with large networks of connections. Those in environments without books, toys, and stimulation have fewer connections.

Why is that important? New information is encoded via existing connections. New connections can be made, and, indeed, are made every day, but the information must travel down existing pathways first. That's why parents and teachers are encouraged to link information being taught to something the children already know. If more connections are available, the child is more likely to be successful in school. That's the primary reason amazing results have been seen from preschool interventions with children who may not have easy access to stimulating environments.

In addition, memories are recalled through pathways in the brain. The more frequently a memory is recalled, the easier it is to be recalled later. Individuals with PTSD (Post Traumatic Stress Disorder) continue to relive their trauma because the pathway to the memories gets worn down like a rutted road and it becomes difficult to get themselves out of it. Therapists are having some success with pairing techniques that essentially pair the trauma with a pleasant memory (another pathway) until the brain can jump out of the rut onto a different road. That's a simplistic explanation, but the connections are the pathways, and they determine how we can use our brains.

Decades ago, children were taught to memorize information. "Drill and kill," or rote memory, seared the information into the children's brains by repeating it over and over until the pathway was easily accessed. Educational researchers found that the children were able to repeat facts but often had little understanding of what the fact meant. Rote memory was labeled as bad and schools began to teach for understanding. The problem there is that children in a classroom of twenty-five students or more don't always develop the pathways to connect memories. Some memorization is good for the brain, but the information needs to be linked to existing pathways to provide meaning.

Connections can be developed throughout the lifespan. Indeed, continuing to learn new material, linking it to stored knowledge, and consciously recalling information from stored memory are the best ways to keep our brains functioning through old age. It won't guarantee dementia won't develop, but it's our best bet.

Electricity and Chemistry

Another basic concept about the brain that needs to be made clear before we go much further is something that many scientists take for granted, but people in the general population don't necessarily think about. The body's nervous system—the system that connects the brain with nerves throughout the body—runs on electricity and chemistry. The electrical impulses that move down a nerve cell at lightning speed are just like the electricity that moves through an electrical cord.

For the impulse to jump to the next nerve cell in line, the electrical energy is changed to chemical energy. The cell with the impulse generates the chemical; the chemical flows to the next nerve cell; and when it is absorbed into the second nerve cell, the chemical energy is converted back to electricity, and the message is on its way. This takes place in the blink of an eye. The chemicals that transmit the signal are called neurotransmitters. They include chemicals such as serotonin, dopamine, norepinephrine, and GABA (Gamma-Aminobutyric Acid, which reduces excitability.)

One common example of how neurotransmitters affect behavior is that individuals with depression often have low concentrations of serotonin. They are often prescribed SSRIs. SSRI means Specific Serotonin Reuptake Inhibitor. Our bodies need serotonin in solution around the nerve endings. A basic description of how the SSRI works is that one nerve cell generates serotonin and sends it to the next cell, then the cell absorbs too much of it. The SSRI essentially plugs some of the areas that would be used to absorb the serotonin, so it will stay in solution longer. The individual's depression is decreased because there is an increase in the serotonin that is left in solution.

By the way, if the nerve cell is encased in a special fat, the electrical impulse travels even faster. The encasement is called the myelin sheath. It develops over time and helps nerves function faster.

The crux of this developmental miracle is that each brain has a unique set of blueprints combined with a unique interaction with the person's environment. Even twins, who have identical DNA, interact differently with their environments. So, with over seven billion people on the earth, each one is amazingly unique. And each one is amazingly important.

If you've ever put together a puzzle only to find the last piece missing from the box, you know the importance of the whole. Every person on the planet has one piece of the seven-billion-piece puzzle, whether they have advanced degrees, drive a garbage truck, or live on the streets. The whole will only really work when each piece of the puzzle fills its space.

I don't do much therapy, but people who come

to me with depression often begin their healing by realizing that they are important. They matter. If the proverbial butterfly flapping its wings in one place can cause a hurricane somewhere else, imagine what people can do.

Interruptions in the electricity and chemistry in our bodies cause difficulties with our functioning, just as having difficulty with the electricity to our houses or businesses can cause difficulties. Again, this is simplistic, but too much electricity can be a seizure, too little can interfere with muscle function (a symptom associated with multiple sclerosis). Chemical irregularities cause problems as well. Too much dopamine is associated with schizophrenia, while too little is associated with Parkinson's. Too little serotonin is associated with depression, and too much norepinephrine is associated with high stress and anxiety. Decades ago, many psychiatric medications just attempted to deaden responses in an effort to decrease the symptoms. Now, however, medications can be prescribed that actually address the chemical issue. Believe it or not, there are still many parents who refuse medications because they don't want to "dope up" their child. Modern physicians don't want to "dope up" the child either. They are working to address chemical imbalances to allow the child to function better.

Strengths and Weaknesses

This discussion leads to why knowing your child's brain is important. Every individual has strengths and weaknesses. Parents and teachers have the responsibility of helping each child find his or her strengths. Children are often painfully aware of their weaknesses, particularly if school does not utilize their area of strength.

For example, if the child's strength is solving a particular problem with the materials at hand, working through prescribed math problems or reading exercises will not help them feel useful. Of course, math and reading are extremely important. I am not advocating for canceling them, but that child would benefit from being involved in an extracurricular team in which students use specific materials for solving problems. They do exist.

People who excel at global thinking—that is, seeing the whole—often make wonderful CEOs because they can wrap their brains around what is going on in the entire operation. That ability is typically associated with people who are "right-brained," which will be discussed later. However, these children are often frustrated in school because they need to see the purpose of the lesson before they start. These are the children who can drive teachers crazy by repeatedly asking, "Why are we doing this?" The well-meaning teacher, who is generally not right-brained, replies,

"Don't worry about that. Just do these steps one at a time, and you will get to the end." What the teacher needs to realize is that it's extremely difficult for global-thinking children to take that first step if they don't know where they are headed. If the teacher could say something like, "We need to know how to compute the area of a rectangle because someday you will need to know, for example, how much carpet to buy to cover the floor in your living room," the child could say "Okay" and get on with it.

Some children learn best by hearing, some by seeing, and most by doing. In general, elementary schools are much better at providing multiple ways to experience learning than middle schools, and middle schools more so than high schools. There are many reasons for this, but those are not the focus of this book. It is painfully true, however. Once I was walking down the hallway in a college education building and heard the professor lecturing to a room full of students about how lecturing is the least effective way to teach. I can't believe he didn't see the irony there.

For example, one mother brought her daughter to me and said that the daughter had difficulty with reading comprehension. She previously had the daughter tested and paid for tutoring, but the girl continued to do poorly on tests for reading comprehension, although not poorly enough to qualify as having a learning disability. That distinction will be discussed at a later time. In many other areas, the girl did just fine.

I tested the girl and found her visual processing to be quite low, but her auditory processing was a strength for her. Reading comprehension tests typically require the student to read silently, then answer questions. On the tests, she could not pass items that were below her age level. Just to see what would happen, I had her read more difficult items out loud and then answer the questions. She was able to answer items above her age level accurately. The problem was not that she didn't understand the information, she just needed to *hear* the information.

We worked with the school to allow her to do "silent reading" away from other students so she could read out loud. She also had the option of listening to books online. Her performance in reading comprehension increased dramatically. It wasn't the reading that was getting in her way, but the *silent* reading.

There are even more students who often can't proceed when the only stimulation is auditory. A parent will say, "Go upstairs, brush your teeth, and put on your pajamas." The child will go upstairs, maybe put on pajamas, but no way will he remember to brush his teeth. Believe it or not, he is not being oppositional. He doesn't remember all three instructions.

This could be a memory issue or an auditory issue. The parent would likely get better results if he had a chart including pictures of a toothbrush and of pajamas. If Dad points to the pictures and says, "I need you to do these things now," the child will likely figure out that he has to go upstairs to do it. Then he only has two things to remember, and with the visual image in mind, he will be more likely to do both tasks.

For the child who excels in the arts, school can be the most boring place on earth. A mother brought her son (whom I'll call Jacob) to me because he was not doing well in reading, and he was starting to act out in school. Jacob hated school, and he was quick to tell me that.

Tests for visual-motor integration (which include tests for visual perception and motor coordination) were given, as I often check visual perception when there are issues with reading. Jacob's ability to draw shapes was amazing. I asked if he liked to draw, and he emphatically said that he did. I allowed him to draw while I conducted tests that were oral in nature. He produced pictures that looked like someone much older had drawn them even though he was listening and answering my questions. He was only eight years old, but he was insightful enough to tell me that drawing calmed him.

When I was explaining the results of the tests

to his mom, she asked if she should schedule Jacob for more tutoring for reading. He was currently receiving an hour a day after school. I suggested that the extra tutoring in reading be canceled in favor of art lessons.

Imagine a task you do poorly. Now, as an adult, imagine being told that because you do it poorly, you need to spend extra time on it every day. Few adults would tolerate it, yet we subject our children to it every day. By celebrating his artistic talent, we helped to validate Jacob. We worked with the school to allow Jacob to draw at his desk when the teacher was addressing the class as long as he could demonstrate that he was listening. He helped the teacher design bulletin boards, and eventually he was even asked to help with some of the art displays in the hallway. He was allowed to spend extra time with the art teacher. Jacob's inappropriate behaviors stopped. And the bonus was that when he felt good about himself, his reading improved as well. It did not become a strength for him, but it did improve.

It is not unusual for a person's weaknesses to improve when they feel better about themselves, although the discussion of self-esteem is fraught with difficulties and strong opinions. Self-esteem will be covered in more detail in Chapter 7.

The intention of this book is to help the reader understand different strengths and weaknesses children can have and how to help them facilitate the development of children's strengths. The readers might even learn something that will help them understand their own brains better.

2—Brain Functions

Strengths and Weaknesses

Chapter 1 presented a brief description of strengths and weaknesses children can have. The purpose of this chapter is to delve a bit deeper.

Our brains perform a myriad of different functions, but the ones most relevant to performance in school are short-term memory (both auditory and visual), long-term memory (auditory and visual), visual and auditory processing, spatial perception, processing speed, verbal ability, and nonverbal ability. Processing speed, verbal ability, and nonverbal ability will be discussed in their own chapters.

Short-term Memory

Short-term memory is essentially remembering something for about twenty seconds. For example, when you look up a phone number and have to re-member it long enough to key it into your phone, that's short-term memory. Some people remember best what they see, some what they hear. Most adults are able to tell me which one they do best. "Oh, I have to see it," or "Just tell me."

Personally, I recall information best when I see it, which might seem odd for a musician. I worked with a conductor once who said, while we were standing backstage, "Oh, let's do this piece."

"I don't know that one."

"Oh. Okay. Your part goes like this—la la la la la…" The first time he did that, I thought he was teas-ing. He wasn't. On the other hand, if I see the music, I remember the part fairly well. If I see it and sing it a couple of times, I have it. In order to remember pas-sages of sacred writings, I have to set them to music. The effect of that is that I don't know that many pas-sages, but the ones I know, I know well.

Someone can give me directions to a place, and I just look at them like they are from Mars. Draw me a map, please. Of course, I know a very intelligent woman who drew a map from the position she was sitting at the table, and north ended up at the bot-tom of the page, but usually that's okay. Just a note: there's nothing wrong with south being at the top. I love maps of the world with south at the top, but the person needing to use the map would have appreciat-ed being informed.

The problem with students who are like me is that teachers talk a lot. Really, a lot. They have too much to explain and are pressured by how much time they have. Sometimes they also talk really fast. The child who needs to see the information is lucky to actually process a small portion of what is said.

On top of that, older students are expected to hear everything, figure out what is important and write it down while they are continuing to listen to more new information. If the students have a learning disability, their parents and case managers can write an Individual Educational Plan (IEP) for them that requires the teacher to furnish notes to the student before class starts. That way they can follow along and hopefully absorb more information. Students without a learning disability are on their own. Most teachers these days do write important points on the Smart Board, though, and that helps.

It would be nice if teachers provided notes for every student. When I give a workshop to teachers, you can be sure they expect a handout with the PowerPoint information on it. They just don't always provide that service to their students. In their defense, they have many more presentations to make. I get frustrated when a teacher, particularly in high school, says, "Note-taking is a skill the students need to develop."

That may be true, but it is much more difficult

for those students who recall information best when they see it. Many colleges of which I am aware now have professors post the PowerPoints for their lectures ahead of time so those students who want it can print out the handout before they come to class. Hopefully, high schools will be transitioning to that process soon.

Students who recall best when they hear information have the advantage when the teacher is lecturing. Where they can stumble is when the teacher provides a handout and the directions are projected on the board, but they are not read by the teacher. It often helps students to read the directions out loud, like the girl mentioned in chapter one. Those students may also need to talk out loud while they process the information as well, although technically, that's auditory processing rather than auditory memory.

Decades ago, William Glasser developed a retention triangle. It stated that in general, people remember 10 percent of what they read, 20 percent of what they hear, 30 percent of what they see, 50 percent of what they hear and see, 70 percent of what they discuss with others, 80 percent of what they experience personally, and 90 percent of what they teach others. The theory has some critics, and it is likely that the nice round numbers were not generated directly from robust research. Nevertheless, if we accept the basic premise, humans remember the least of what they hear and read. How are students taught most frequently? Through reading and lecture. Scary.

Visual and Auditory Processing

Processing information is not quite the same as recalling it, although it is linked. With short-term memory, information is there and gone unless something else happens. In order for a child to recall the information later, it must be processed.

Many of the difficulties described with visual and auditory short-term memory also apply to visual and auditory processing. The strategies work as well. People won't remember information unless they consciously process it. Then they need to connect the new information with something they already know. They can organize it and file it, but if a link is not made, that file is going to be difficult to find when they need the information.

For example, imagine being given an important piece of paper to keep track of. You put it into a file folder and into a file drawer. A year passes. You need to find that piece of paper. Without a label on the file drawer and the file folder, that task is nearly impossible. The labeling is the *link*. The drawer could be, "Important papers about the mortgage," and the file folder could be "receipt for the extra payment sent to my mortgage company." With the labels, the paper is easy to find.

Your brain is sort of like a file cabinet, although thinking of a spider web might be more accurate. One idea about mortgages can be found by thinking about mortgages in general, then to the memory of the workshop you went to about making one extra payment a year. You picture yourself at the workshop, recall your resolve, then come up with, "Oh, yes! I did send an extra payment last year!" That's *processing* the memory.

Students who learn best by seeing can sometimes use different colors as they take notes, use different types of handwriting, or draw small pictures to help clarify what the teacher is saying. When studying for a test, it often helps them to put information on colored note cards, arranging those concepts that are linked to the same topic by color as well.

I used to use one of those pens that had different colors of ink. I would switch handwriting and ink colors, then during a test, a question would trigger a memory of the green notes I had made on the right-hand side of my notebook.

Students who learn best by hearing can record lectures then refer to passages from the recordings later when they are organizing data. They can associate different facts with different melodies. Melodies, not songs. Songs with words will likely interfere with the information the person is attempting to store. Chanting the information in a rhythm often works very well, as our brains naturally respond to the patterns. Dementia patients, who have difficulty with memory, can often recall songs and playground chants long past the time when they can carry on a conversation. I hope I remember the pleasant ones and not the ones we made up when the teacher wasn't around, though.

Long-term Memory

Long-term memory is technically anything one remembers longer than twenty seconds. It includes remembering what you had for lunch today as well as memories from elementary school. The more often a memory is recalled, the easier it will be to recall again. As mentioned in chapter one, this sometimes works against us when we recall bad memories more and more often. Therapists have had some success in treating PTSD by systematically pairing pleasant memories with the bad ones to provide an alternative pathway. Sometimes, this requires the addition of an alternative outcome. For example, for rape victims, trained therapists are learning to help the victims generate an outcome in which they overpower the rapist, and the rape does not occur. They aren't erasing the memory, but the trauma of the rape develops one of those ruts I discussed earlier. The alternative ending provides a different path in an effort to lessen the trauma.

For children, it helps them tremendously when

a parent or caregiver asks them about their day when they get home from school. Ask them some specifics about what they learned and how they felt about it. What was the most interesting part? What was the funniest part? Try not to ask yes or no questions. They encourage one-word answers, and that's not the goal. This process will help the child create a pathway that has been used more than just the initial time it occurred. The bonus is that it also nurtures the relationship between the adult and the child.

Children are also more likely to take time to learn/memorize information when the parents demonstrate that it is important to them as well. Children naturally seek to emulate their parents. If you teach your child that she shouldn't lie, then tell a lie to someone else in front of her, she is going to learn that lying is okay. The old adage "do as I say, not as I do" is nearly impossible for children to follow. They will do what their parents do almost every time.

Parents who demonstrate the importance of memorization to their children also get benefits. When they memorize information, their own memories are strengthened. The saying, "use it or lose it" applies here. If you want a good memory as you age, work on your memory now.

The visual processing and memory strategies work well for long-term memory. The auditory strategies of listening again to a lecture, reading out loud, and rhythmic chants work for auditory learners. The best results come from using both visual and auditory strategies as much as possible because each aspect of the strategy is a potential pathway for recalling memories.

One thing we need to realize is that our memories are not infallible, particularly when there is emotion involved. There was an amazing study done just after the Challenger space shuttle disaster occurred. A psychology professor asked his freshmen students to answer a few questions about what they were doing when they heard the news. When they were seniors, he asked as many as he could find to complete the same questions, adding, "How sure are you that your memory is correct?"

Nearly 100% of the students answered that they were sure they were answering the questions exactly the same way they had when they were freshmen. Fewer than 30% actually matched responses. Now, my memory about reading this study years ago told me that it was fewer than 10%, but I'm allowing for misinformation over time.

Our memories are directly linked to our perceptions of self. It's difficult to realize that they might not always be correct. I once saw a poster in a social worker's office at one of the schools in which I was working at the time. It read: "There's your truth, my truth, and the actual truth." So true. Sometime when you have a few minutes, listen to the song, "I Remember It Well" from the musical *Gigi*. It's hilarious.

* * *

Long-term memory directly links to another topic that is critical in brain function. Students can organize information by color or melody, have the topics clearly in mind, and still struggle when they get to the test. One primary reason is that memories are converted to long-term storage during sleep. But it has to be enough sleep and of good quality.

As a society, the United States is one of the most sleep-deprived nations in the world. Both Mayo Clinic and the National Sleep Foundation state that elementary students need nine to eleven hours of sleep each night, and teenagers need eight to ten.

I have worked with students for decades, and I have seldom seen students who get enough sleep. That may have to be taken with a grain of salt, as I also seldom see students who don't have learning problems or mental health issues, but that says something in and of itself. But which is the chicken and which is the egg?

One student I had that was struggling in school was living with her grandparents, and she slept on the couch. That was not the problem. The problem was that the grandparents watched television in the living room until late at night, and only then could the eight-year-old girl go to bed. The school gently encouraged the grandparents to move the television into their bedroom so the girl could get more sleep. After they moved the television, suddenly the child started doing better in school.

A mother who brought her daughter to me wondered if the girl had Attention Deficit Hyperactivity Disorder (ADHD). The child was out of control. She could not settle down or be comforted. I asked the mom how much sleep the child got. Mom said, "Oh, she never sleeps more than three or four hours a day." The girl was three. A three-to-five-year old needs between ten and thirteen hours of sleep a night. I told the mother to make sure the child had no caffeine, provided her with several relaxing techniques, recommended a weighted blanket, and told her to bring the girl back in three to four months. The poor mother, who was understandably at her wits' end, asked what she could do if those techniques didn't work. I recommended asking her physician for help by prescribing a medication that would help the child sleep. I should add that the child had been adopted and was drug positive at birth.

Prescriptions are rarely needed for a three-year-old, but this was a critical case. Lack of sleep is a form of torture. Adults who are deprived of sleep for days often struggle to concentrate, can hallucinate, and can experience psychotic episodes. Lack of sleep is also cumulative. This poor child was so sleep-deprived, she could not control her actions to the extent that a typical three-year-old does. The mother called two months later and said that her daughter was doing better. Her behaviors were not yet under control, but she was sleeping more. The physician had recommended a small dose of Benadryl, and that had started the sleep process. After several days, she started sleeping more on her own. Apparently, she had been so exhausted that she couldn't get to sleep.

The girl also benefitted tremendously from having a weighted blanket. Weighted blankets were used decades ago to help calm students with autism in the schools. Teachers then found that they helped with students with sensory processing disorders such as ADHD, depression and anxiety. I have one in my office, and part of the intake process typically includes having first the parent then the child lie on the couch in my waiting room. I put the fifteen-pound blanket on top of them, and 98 percent of them relax visibly in front of me. Sometimes the parents are reluctant to give the children a turn.

The one parent who didn't like it at all was an EMT. He said, "I could not get out of it quick enough." That's true, but I felt a little sad that he never allowed himself to really relax. He wasn't my patient, though, so I didn't say anything.

The various schools where I have worked have had weighted blankets. Many of them also have weighted vests, but these are not used unless an occupational therapist is monitoring its use because the weight of the vest can affect the child's posture and spine.

Spatial Perception

Do you know people who just have a knack for seeing how things fit together? Their outfits always look just right, and their homes look like something out of a magazine. I'm not one of those people. I can, however, appreciate when something does look good. I just can't get it to look that way. On the other hand, I have a daughter who can do a thousand-piece puzzle in two or three hours. She can look into a heap of pieces and pull out the one that goes in a particular spot.

It's true that this group of people typically include artists, and they make the world more beautiful for the rest of us. However, there is more to spatial perception than art. Although we use words as our primary avenue of communication, most communication, in reality, takes place nonverbally. Facial expressions and body language communicate much more effectively, much more honestly than words.

A person's true reaction to a comment or situation is usually reflected in their facial expression—fleetingly—before they recover and put on the face that is more acceptable. Those who are good at reading people are good at noticing those initial responses. Actors and actresses are very talented at prolonging those initial responses a bit so the audience can pick up on them.

Unfortunately, when a typical person reads someone else's facial expression, they process the information through their own brain. What other option do they have? But that often plays with the accuracy of the interpretation. For example, when I was writing down information during an interview of a new patient, she told me something, and I frowned. She immediately interpreted that as my perceiving that there was something wrong with her. She said, "What is it? What do I have?"

After processing that she was referring to my frown, I answered, "I was frowning because I don't know how to spell that medication."

I confess that when I left graduate school, my committee chair told me that I was extremely qualified except that I needed to work on my poker face. I don't have a poker face. Neither did my father, which makes me recall a lot of fond memories about playing games as a family. We always knew when he had the bad card. I also spent a lot of time on stage in my life before graduate studies. Let's just say I have an expressive face. The only way I get around it while testing is to always smile and be encouraging whether the client is doing well or not. Nevertheless, I believe that my expressive face often helps me when working with children. I do try to let people know what I'm thinking when a frown crosses my face though. How does this affect children?

First of all, when Suzie sees Mandy looking at her, she might believe that Mandy is saying something bad about her or doesn't like her. Mandy could just be looking in Suzie's *direction*, not looking *at* Suzie at all, but Suzie might get angry, get hurt, and maybe even start an argument with Mandy at recess. And Mandy tells the teacher, "I didn't *do* anything!" Yet, in Suzie's perception, Mandy did.

With one glance, a friendship may change, a day can be ruined, and—this is key—Suzie most likely did not learn anything during the time she was upset.

For the majority of children, the development of the ability to read facial expressions happens without any direct instruction. However, the interpretation goes beyond identifying the emotion being expressed. For another example: Mandy could see Suzie frown, and while correctly interpreting that Suzie was not happy about something, she could incorrectly interpret that whatever Suzie was not happy about is related to her (Mandy). Because children are, by nature of their development, egocentric, everything in their world is about them. Of course, most adults retain this in some part of their characters.

It can be helpful to discuss the meaning of expressions with children. Making it a proactive, conscious task rather than a reactive task can help them learn that everyone has emotions and thoughts that are displayed on their faces. However, seldom does another person, particularly a child, truly understand what that facial expression means.

Another example: Davion asks his teacher a question, but at the same moment, his teacher experiences a pain in her foot because of an ill-fitting shoe. Davion immediately interprets her frown as indicating that his teacher doesn't like him or thinks his question is stupid—and so is he.

Some readers might believe that I am over exaggerating, but this small incident can lead to Davion acting out in class or becoming sad while believing he must be stupid. I've seen it hundreds of times.

One of the important lessons parents can teach their children is that frowns or other facial expressions aren't always about them. The same goes for body language, although younger children are not often aware of body language as much as older children and teens.

The teen years involve the perfection of body language. Go into a high school classroom, and you will be able to read those students who are portraying boredom, those who are appearing interested, and those whose thoughts are elsewhere. You will also see students copying what their peers are doing so they fit in. Adults are a bit sneakier. We are more aware of the appropriateness of body language, so we work to appear interested when it is expected even if we are extremely bored.

With slumped posture, eye rolls, and bored expressions, teenagers can often frustrate the adults around them. Parents and teachers know how rude the behaviors are and frequently get offended. Often, however, teens are knowingly presenting the body language to test the adults to see if they will take the effort to get to know the person beyond the behaviors.

The behaviors of teenagers could take up an entire book by themselves, but the purpose of this book is to focus on their brains. Therefore, try to remember, when teenagers are driving you crazy, that their brains are going through significant changes right along with their bodies. Remember that growth spurt of connections which later get trimmed by killing off brain cells discussed in chapter one? Take a moment to recall your own teen years and try to get some perspective.

Brief tangent: an important message related to perspective is that people cannot control the actions of others, but only their own response to that action. (This is looking at the broad picture without getting into a discussion of the responsibility of parents and teachers to control their children and students.) I often provide adults with stories to clarify understanding. If you've ever been driving and had a driver pull in front of you, almost clipping your front fender, you know how infuriating that can be. You have two choices. You can scream at that person, get angry, and possibly let it ruin your day by carrying the frustration around with you. Or you can assume that he has a family emergency and say a prayer for him. Neither one of those choices is going to change his day, with the possible exception that the prayer may help him, depending on your beliefs. On the other hand, if you select the second option, *your* day will be better. You did something nice for someone else and kept anger from getting control of your brain. Try it sometime. There are times that I forget, of course, but I'm working on it.

One convenient way to discuss facial expressions and body language can occur while watching a movie with your children. Stop the action and actually talk about what the characters' faces are communicating. When they get good at it, you can use a subtler scene and ask the children if the character is lying. Perhaps you could make a game of it, predicting what the character in the movie was going to do by reading body language and facial expressions. Mute the dialogue. Mozart made it easy. In his operas, if the character was sad but the music was happy, the character was lying. The music was always right. It's not that easy in real life.

Individuals who do well at spatial perception are often good at art, computers, design, and understanding what others are feeling from reading body language. Notice that reading and math are not necessarily on this list. If this is your child's area of strength, it might help to offset frustrations that may occur in school with such activities as extra art lessons or robot-building clubs. These children might also do very well on teams focusing on mediation. Many schools have teams of students that are trained in mediation to help settle conflicts between students, and it makes sense that students who excel in reading body language and facial expressions would be very good at it.

Tressa Reisetter PhD, LP, NCSP

3—Processing Speed

The speed at which a person processes information has historically been equated with intelligence. When a person is described as quick, he is considered smart. When he is described as slow, he is *not* considered smart. More recent brain research has further clarified that there are many tasks that the brain does that can affect its ability to function well. Nevertheless, processing speed is one of the most important tasks, as it affects other aspects of intelligence.

* * *

My dissertation was on processing speed. I analyzed data from over two thousand patients my neuropsychology professor, Dr. Ray Dean, had evaluated over several years (no names of the patients were included). The patients had diagnoses ranging from brain tumors, to learning disabilities, to traumatic brain injuries (TBIs), to ADHD. They were given over twenty tests from the *Woodcock-Johnson Tests of Cognitive Ability* (editions R and III), and for over 90 percent of them, their lowest score was processing speed. In other words, no matter what diagnosis the individuals had, when there was a problem with the function of their brains, it affected processing speed the most.

It makes sense, then, that many neuropsychological tests have a timed component. Companies that produce tests to measure aspects of intelligence are careful to not include a timed element when they are attempting to measure a different aspect such as verbal ability. It is important not to muddy the waters.

One reason I find processing speed interesting is because I am a musician. For decades, students who were in band or orchestra averaged about thirty points higher on *both* the verbal and math parts of the SAT than students who were not in instrumental groups. At first it was assumed that students from more affluent families were in band or orchestra, and affluence was the primary difference. Students from affluent families perform better on the SAT. However, when socioeconomic status was controlled by statistical analysis, the music students still out performed their non-musical peers. My husband, who is also a musician, and I suspect that it was related to processing speed. Music takes place in real time.

Physical therapists or personal trainers get more out of your muscles than you would get on your own because they push you past the point where you would stop if they weren't watching. Likewise, a conductor makes the individuals in her band or orchestra play faster than they would play by themselves. When the students are practicing, they subconsciously slow down when the musical passages are more difficult to play.

Choral students often performed better on the SAT than the overall average, but not as much as a group compared to instrumental students. One likely explanation is that many choral teachers teach their students by rote rather than making them read the music. Apparently, it's the reading music and playing simultaneously that affects performance on the SAT tests, which are timed.

One of the interesting things I discovered about processing speed when I was researching my dissertation was that tests of processing speed could differentiate between children with and without learning disabilities in reading. Back in the 1970s and 1980s, researchers administered a rapid naming test, and the children with a learning disability in reading scored significantly lower than children with no learning disability and with disabilities in other subjects.

In the 1990s, other researchers posited that pro-

cessing speed was so important it should be a separate area of learning disability. Researchers also
found that processing speed measured in first- and
second-graders was the single best predictor of reading success in middle school.

It was documented that experience could affect
processing speed. Retaking a timed test often results in a better score. It's called the practice effect.
But the effect is broader than that. For example, a
twelve-year-old generally has a slower processing
speed than a twenty-year-old. But a twelve-year-old
experienced in playing chess can scan a chessboard
more quickly than a twenty-year-old who has no experience in chess. Experience counts.

A study that intrigued me was one in which individuals with learning disabilities in reading scored
significantly lower than individuals with no reading
disability. The task was simply to keep a steady beat
along with a metronome. The trick was that the individuals had to use alternating index fingers. In other
words, they tapped right-left-right-left along with the
beat of the metronome. The individuals with disabilities in reading had significant difficulties with this
task, even though there was no reading involved.

With the exception of the metronome test, most
researchers I found used a test for process speed that
involved a task usually associated with reading itself,
but always with a task that had to be taught such as

naming colors or numbers. A task that has to be taught is generally associated with academic achievement rather than cognitive ability. The research connected to my dissertation found that a cognitive measure of processing speed (from a cognitive test battery rather than an achievement battery) distinguished between individuals with and without learning disabilities. The results supported the supposition that the difficulty individuals with reading disabilities have with speed is not confined to their ability to identify learned concepts but is linked to the processing of information.

Special Education Programs

My first job after graduate school was in a school district that specialized in working with students in a setting for special education programs. Special education in Minnesota essentially has four settings. If students are Setting 1, they remain in the general classroom all day with specific accommodations because of their defined disability area. For Setting 2, students spend the majority of their school day in the general education classroom, but they go to a resource room for special instruction in their disability area. Setting 3 students spend most of their day in a separate classroom and only join their general education peers for situations like PE, music, or lunch.

Students in a Setting 4 program attend classes in a different building than their general education peers. Most of these students have severe behaviors linked to Autism or to unknown causes. The students without autism who demonstrate significant behaviors that interrupt their own education or the education of others are typically serviced under the category of Emotional or Behavioral Disabilities (EBD). Different states may have different names for the category.

The thing that frustrates me about students in the EBD category is that too many adults, both parents and teachers, believe that the behaviors the students are exhibiting are willful. The whole concept of a disability is that it is not willful. Of course, every student acts out willfully sometimes. However, too many adults who work with these children believe they are just "bad kids."

As I said in the first chapter, I believe the vast majority of students do not misbehave on purpose for the majority of times they misbehave. When I tested these Setting 4 students, 80 percent of them had a significant weakness in some area of cognitive functioning and 50 percent of them had a significant weakness in processing speed. Their scores provided evidence that they were not in total control of their behaviors.

Weaknesses in Processing Speed

What symptoms do students exhibit when they have

a weakness in processing speed? For one thing, do you remember those fleeting facial expressions discussed in Chapter 2? Individuals with slow processing speed usually miss them altogether. If a person's true reaction to given information is expressed in those first microseconds, the person with a weakness in processing speed will not understand what that person is thinking or feeling. That can often be the foundation of miscommunication.

Another symptom is related to the fact that our brains do not like gaps in information. If you've ever had someone tell you something as you're going out the door and you miss part of it, you understand this. You're not exactly sure what the person said. You could go back in and ask, but you're in a hurry and don't want to do that. So you repeat what you did understand, and from that you figure out what the person must have said. Often you are correct, but not always.

For the person with a weakness in processing speed, however, normal conversation sounds like a taped conversation played back at a faster speed sounds to us. They miss chunks of conversations or lectures by teachers. There are too many chunks for them to analyze each one, so their brains adapt by putting in information that makes sense. It may have little to do with what was actually said.

For example, in the morning, Omar's mother tells him that he needs to be home by three o'clock because she has a doctor appointment at four o'clock. What Omar remembers is, "I have a doctor appointment at three o'clock, so you don't have to be home until four o'clock." When she finds him playing with friends at three-thirty, she is really angry.

"I told you to be home at three!"

"No you didn't! You said you had a doctor appointment at three!"

"No I didn't!" She's right, of course, but what she doesn't realize is that Omar is telling her what he remembers even though it's not what she said. She adds further damage when she says, "You are lying!"

Students with slow processing speed get labeled as liars or as being oppositional because they don't do what they are told and then argue about it. If the students can recognize that their memories can be faulty, and parents and teachers recognize that the student is very likely saying what he *remembers,* a lot of grief can be avoided.

When working with these students, parents and teachers need to make sure important information is repeated, or, better yet, written down for the child to refer to later. For young children, visual charts with pictures of tasks to be done can help tremendously. For older children, class notes need to be furnished by the teacher before class, so the student can follow along as best as he can (as mentioned previously).

In school, students with slow processing speed

can be overwhelmed by mid-morning. Everything is moving so fast that they feel lost. Everyone around them seems to be doing just fine, so they start to believe that they are stupid. When they feel that they cannot handle what the other children seem to handle with no problem, they often do one of two things. They either shut down, internalize the feelings and start down the road to depression, or they start acting out. Adults may say to me, "But you said she wouldn't act out on purpose."

What's important to realize is that these students are too young to analyze what they are doing. The *need* is to divert attention from their problem so no one else will start to think that they are stupid. The *action* of misbehaving (bothering their neighbor, making noise, leaving their seat) stems from a need that can be addressed, not from the child being bad.

I honestly don't understand why, but over decades of working in schools, I have come to realize that students would rather be considered "bad" than "stupid." In my family, badness would be much worse than stupidity. On the other hand, I realize that's easy to say since I did well in school. Perhaps it has something to do with society's glorification of bad behavior. We have some sort of fascination with outlaws (Jesse James, Bonnie and Clyde, etc.), but movies about people who are less cognitively endowed than average are more like *Dumb and Dumber.* We laugh at them. Maybe that's why children don't want to be considered dumb.

What Can Be Done About It?

A couple of years ago, I printed out a document from the Minnesota Department of Education (MDE) that provided lists of activities to improve various areas of cognitive functioning. Cognitive test scores are not carved in stone, so there's always something people can do to improve their performance. The list from MDE included things people could do to improve verbal ability, spatial ability, short-term memory, long-term memory, etc., but it actually stated that there was nothing that could be done to improve processing speed. I completely disagree with that!

First of all, recall the music students who earned higher scores on the SAT. Clearly, *something* had helped them improve both verbally and mathematically, and since other factors were mitigated by statistical analysis, it had something to do with music. I believe the music increased their processing speed.

To test that theory, my husband and I set up a research group after school with seventh grade girls (it was not a purposeful part of the research to have only girls). The research went on over a four-week period where the girls were divided into three groups.

If anyone takes a timed test a second time, we would expect their score to increase slightly because they are more familiar with the test. Therefore, one of the groups only came in to take the test on the first and last day.

To rule out any improvement that might be associated with more time with an adult, the second group came in after school two days each week and played games like Scrabble, Scattegories, and Taboo with a second teacher.

The third group did music activities with my husband on the same two days after school. The activities were not really different than many activities that take place in general music classes all over the country except that they worked on speed. For example, they read rhythms, then worked on going faster and faster. They sang songs with a lot of words and worked on going faster. They also did patterns of movements and worked on going faster. Music teachers do not concentrate on increasing the speed of the various activities. They are focused on teaching the music curricula.

The tasks for our research were aimed at increasing processing speed, rather than teaching music. They are two different things.

At the end of the four weeks, the girls all took the test again. There was a slight improvement with the first group, as expected, but it was not statistically significant. The same was true for the second group. The third group improved significantly as a whole. An additional, unexpected outcome was that we had all of the girls come in two months later to take the test to see if the effect was lost when they no longer participated in the after-school groups. The first two groups stayed the same, but the third group improved again. The gain was smaller than when they were actively participating in the music group, but nevertheless, there was further improvement.

That was surprising.

Since processing speed develops along a curve, the students had apparently jumped to a higher curve. Additional results would have to be done to confirm this, of course. We do know, however, that individuals with learning disabilities develop processing speed along a shorter curve than individuals without learning disabilities, so different curves do exist.

* * *

The curve for developing processing speed is steepest when a child is six to eight years old, in approximately first and second grade. That means that during those years, the window for developing a fast processing speed is open wider.

With that in mind, my husband and I talked to a couple of elementary school music teachers, and they agreed to spend about ten minutes of their time with first and second grade students to do the activities that were used in the project with seventh grade girls. An interesting aspect of this research was that one of the teachers taught in an affluent school and the other in a school where most of the students were low income.

We got permission from parents to test students before and after the two-month window we were using, and one hundred students signed up.

After the first test administrations, the students in the more affluent school averaged scores of about 102, with the average range between 90 and 110. The students in the low-income school averaged about 94, still in the average range, but a bit lower. This was not unexpected because so many children from low-income situations have early childhood environments that are not necessarily rich in developing brain connections, as discussed earlier.

The amazing result at the end of two months showed the first group's average improved to approximately 110, as did the second group. As you can see, the children from the lower SES caught up with the students from the higher SES!

Clearly, more research needs to be done, but these results were quite promising on several levels.

* * *

When there are difficulties with processing speed, I encourage clients to consciously work on doing tasks faster and faster. Tasks could be a computer game of solitaire, a basic dance move, reading a light work of fiction, and of course, playing a musical instrument.

As long as it's a fairly easy task or a piece of music easier than typically played, you can work on increasing your child's processing speed as well as your own.

4—Verbal Ability

Over a lifetime, the most consistent ability tends to be a person's verbal ability. Of course, when dealing with humans, there are always exceptions to any generalization. Typically, however, when a person is injured or develops an illness that affects functioning, the verbal ability is the best way to predict what that person could do before the injury.

People have gotten frustrated with me because they come in for an assessment after a car accident and want to prove that the accident affected them negatively. This could work well for them if they are trying to sue for damages, but without evidence that they suffered a change in function in that specific time-period, there is no proof.

In a perfect world, accidents would occur right after a neuropsychological assessment was done. That way, it could be said with a fair amount of confidence that any change in cognitive functioning could be attributed to the accident, but there would always be doubt.

Language in the Brain

If someone asked me to predict how well a child would do in school using minimal testing, I would administer a verbal test and a test for processing speed. Children who do well on these two tests will very likely do well in school academically (there is no link with social skills, however).

Humans are not directly wired to read, but they are very strongly wired to communicate. Even primitive humans thousands and thousands of years ago developed ways to communicate. The ability to communicate enables us to build intricate social structures and accomplish amazing feats. I saw a film several years ago describing the building of the Baha'i House of Worship in Delhi, India. There was not access to modern construction equipment, and the workers spoke many different languages, but an architectural wonder resulted. Communication is a basic human need.

For most people, language is located in the left hemisphere of the brain, generally above the left ear.

There are certainly aspects of language elsewhere, but for these purposes, that's the easiest way to think of it. For 99.9 percent of right-handed people, language is on the left side. The left side of the brain runs the right side of the body. Therefore, language on the left side and right-handedness generally coincide.

Two-thirds of left-handed people also have language in the left side of their brains. This combination doesn't always work as naturally for language. The right side of the brain controls the left side of the body, so sometimes there isn't a smooth communication between language on the left and handedness on the left. For these individuals, the control of the left hand comes from the right side of the brain. A dominant right side of the brain is often associated with strengths in spatial perception (possible examples: art, architecture, computers, design, math), with less likelihood of verbally-based strengths. As I've said, these are generalizations. There are always exceptions.

For one-third of left-handed people, language develops on the right side of the brain. I don't know why, and I would love to read a study of how these individuals develop language. Development for the majority of people puts language on the left side. So what happens for the development of these folks that makes language shift to the right side? When I test left-handed people, I can't tell which half of the brain they are using for language. Knowing that the brain seems to work best wired a certain way, however, I would predict that there are some inherent difficulties.

The need for language is so strong that when there is an injury on the left side of the brain, language has been shown to move over and develop in the right hemisphere. Generally, it doesn't work as well over there, but the brain seems willing to sacrifice some of the abilities associated with the right hemisphere to keep language functioning. These individuals can learn to communicate and can even live independently. The brain is extremely adaptable, and there has been research showing that the spatial abilities in these cases can move over to the left hemisphere, but brain function is often more laborious for these individuals, and they often have some type of language disability.

The left side of the brain is so strongly set up for language that when surgeons have put a shunt in to drain liquid off the brain, they typically put it on the right side. That way, if the shunt interferes with brain function, it will be less likely to affect language.

The interesting thing about verbal ability being such a good predictor of success in school is that it is an area parents and care-givers can help a child develop. I realize how difficult parenting is and how pressed parents are for time, but if parents want their children to have the best chance for success in school, they need to read to their children frequently, hopefully every day.

Language development

When children are born, their hearing is the most developed sense they have. There was a study that was done in a hospital nursery. When babies heard another baby cry, they were likely to cry as well. Many people know this. Watch the babies in the nursery, and you can see it happen. The babies in the study even cried when a recording of crying babies was played. They did not cry, however, when the sound of a baby monkey crying was played, and they did not cry when the sound of an adult crying was played. While it is a bit humorous that newborn babies can set each other off, the main takeaway was that the newborn babies could tell the difference between recordings of human babies, monkey babies, and human adults. Amazing!

The cooing and babbling noises babies make are all pretty much the same until approximately six months of age. By that time, the noises the babies make are language specific. I remember reading studies that found those results when I was in a graduate school child development class. A baby in China will sound different than a baby in Russia or the United States after the age of six months.

Children with larger vocabularies do better in

school, as shown by longitudinal studies performed over the course of several years. Children in environments that do not foster language development have fewer words in their vocabulary. When parents read to their children, and work to provide that enriching environment discussed in the first chapter, the children have a larger vocabulary.

Why is that important? The US Department of Education has a report on their website from the Reading First National Conference in 2008 that makes a reference to a study done by Tabor in 2001 of a longitudinal study that found that the size of a preschooler's vocabulary correlated with reading comprehension in upper elementary school. In 1998, Scarborough discovered that the vocabulary size of kindergarten students was "an effective predictor of reading comprehension in middle elementary years." Cunningham and Stanovich found that vocabulary of first grade students predicted reading comprehension ten years later. There was also a study by Chall, Jacobs, and Baldwin that found that children with a smaller vocabulary in third grade scored lower in reading comprehension two to three years later.

The implications of that research indicated that if parents help their toddlers and preschool-aged children develop a larger vocabulary by talking to them and reading to them, those children are more likely to be good readers and have success in school.

It is also important not to discount the effect of parenting by example. Children copy what they see their parents doing. If your children see you reading, they are very likely to development a habit of reading. If parents support the importance of education, their children will more likely believe school is important and strive to do their best.

A study by Norbert Schady in 2011, titled "Parents' Education, Mothers' Vocabulary, and Cognitive Development in Early Childhood: Longitudinal Evidence from Ecuador," found that the strongest predictor of cognitive development in children was the vocabulary and educational level of their mothers. Schady goes on to say that "[h]ousehold wealth and child's height, weight, and hemoglobin levels explained only a modest fraction of the observed associations. The vocabulary levels of mothers and children were more strongly correlated among older children in the sample, suggesting that the effects of a richer maternal vocabulary are cumulative."

I hesitate to bring in this study because it focuses on mothers rather than parents, but it is just too important to leave out. Keep in mind that this study is only one of many that produced similar results.

For a long time, people have assumed that children do well in school in proportion to how much money their parents have. For example, they believe that rich people have children who are successful in school, and the children of parents without much money struggle in school. This study does not support that money is so important.

What that boils down to is that the best way to ensure the success of your child in school is to increase your own vocabulary and share your knowledge with your child. The study was done with mothers and children in Ecuador, but the same holds true in the USA.

I bring this up for two reasons. First, the role of women in the success of their children has been so widely documented that, even back in the 1990s, the International Monetary Fund and the World Bank worked from the premise that the single most important factor in developing impoverished localities was to educate the women and girls. I was teaching a combination Earth Science and Geography class at the time, and the World Bank circulated a movie stressing the importance of teaching women and girls globally. In patriarchal societies, this is not an easy sell. Men need to be persuaded to invest in women for the good of their community. In 2018, this is still a struggle.

The other reason is that even in our developed society, the role of women continues to be cast as subordinate to men. The amount of money women earn for similar jobs is not equal to that of men, but that is improving. The role of women who stay home to raise children is often not as well respected as the role of women who work outside the home. If we consider the future of humanity in general and our own children specifically, we would be smart to focus on the education and elevation of women and girls so that if they become mothers, their children will be more likely to experience success in school and in their careers. I think it's wonderful if fathers want to stay home with their children, reading to them and aiding their development. Research is often aimed at mothers, as they are more likely to be the parent staying home with the child. Traditions can change, however. The important thing is to talk with your children and read to them.

English

I am often thankful that English is my first language. It's a terrible language to learn. There are more exceptions than rules, and I admire anyone who can learn it as a second or third language. Since I only know one language, I am impressed by anyone who speaks English with an accent because I know they speak more languages than I do!

Elementary schools tend to teach English using phonics. There has been a debate between phonics and whole language for years, but currently (at least in Minnesota) phonics is a primary teaching method. If your children struggle, before you get frustrated with them, allow me to give a few examples.

How would you pronounce "ghoti"?

The key is to take the "gh" sound from the word "tough," the "o" sound from the word "women," and the "ti" from the word "action." Ghoti is then pronounced as "fish."

Consider the words tough, though, thought, and through. They should all have the same vowel sounds, right? But they don't.

When there are two consonants, divide between them, and the vowel is short. The rules of phonics would have "dessert" pronounced as "desert" and vice versa.

There are thousands of other examples, but you get the idea. Perhaps focusing on oral language and communication would help develop language skills in young children who are not developmentally ready for the rules of English. While I was doing my dissertation, I read studies about European education. Countries with languages that are more consistent than English, such and Spain or Italy, had a significantly lower percentage of learning disabilities in reading than the United States or England. English is difficult.

We can't jump off the phonics band wagon entirely, however. A child's phonetic skills are excellent predictors of reading success. It relates back to the fact that we are hard-wired to hear and understand.

Hearing is the most developed sense at birth, and it is the first way that babies start to make sense of the outside world. Perhaps a balanced approach is best.

Reading

The phonics programs mentioned above are intended to help children learn to read. Reading is important. In addition to the importance in the subject area of English, reading is necessary to learn other subjects such as history and social studies. Yet, as I mentioned earlier, our brains are not as clearly wired for reading as they are for oral language.

A difficulty connected to this is that schools are teaching reading skills earlier and earlier. In Minnesota, students in first grade typically learn what students in second grade are learning in most other states. Students in kindergarten are reading. Eons ago, when I was in school, the process of reading was not initiated until first grade. In European countries (at least as recently as 2000), they don't even start to teach reading until a child is eight years old, when a child is more developed in the basic skills that support reading.

Years ago, a popular poster titled "Everything I need to know, I learned in kindergarten," was created by Robert Fulghum. It does not mention reading. Perhaps if we focused more on manners and good behavior in kindergarten, there would be fewer disruptions later when we are teaching reading skills. It's just a thought. I do know that classroom management is a huge part of teaching, and it makes sense to me that if the teacher could focus more on content, children would have more content to learn.

Realistically, reading opens the doors to so many worlds I can't begin to imagine what it would be like to not know how to read. When we start teaching the foundations of reading before many children are developmentally ready, however, those who can't make their brains do what is being asked fall behind. It would be like requiring a person who is four feet tall to slam dunk a basketball before advancing to the next grade. They physically cannot do it.

When they can't, though, they know they are falling behind, and their self-esteem plummets. Is it significant that the state of Minnesota consistently is in the top five states for scores on the ACT and SAT, but there is a significant lag for low-income households? On the Minnesota Department of Education website, there are charts showing that the general population of students score average a score of 69.43 percent on the MCA tests administered to school-age children.

Children who qualify for free or reduced lunch (low-income), however, average a score of 38.76 percent. The gap seems to indicate that children in households with a higher income develop faster and can better handle the reading curriculum expectations. What if, instead, the difference is more related to increased exposure to vocabulary and reading done by parents? That is the idea behind preschool programs for children in homes with low incomes, but the general research supports that helping low income parents increase their own vocabularies and those of their children might be able to help close the gap.

Recall what was said earlier about having enriching environments. If impoverished families struggle to survive and don't have time or energy to offer their children experiences that create brain connections, their children are not going to be ready to learn at school. The children often develop low self-esteem, as mentioned above, and the downward spiral begins.

When students come to me with reading disabilities, the first thing I do when I go over testing results is to stress the things they do well. It is always best to build on strengths. Next, I emphasize the importance of using as many senses as possible. If students can hear the story being read while they read along, they are more likely to comprehend. If parents are not good readers, it is perfectly acceptable to use books read by CDs or computers. The libraries have access to hundreds of them!

The Left Hemisphere

Although generalizations oversimplify, there are really some things about the left-brain vs. right-brain construct that warrant attention. The left-brained person is typically strong in language. They organize information in orderly, step-by-step processes.

Recall the outlines teachers used to help students organize their thoughts. There were main headings, sub-points, and further sub-points. This organization helps students write strong papers. It works. It works for most of the students, and when there are twenty-five to thirty students in the class the teacher cannot address the needs of the students for whom it does not work. These students will be discussed in the next chapter.

Left-brained people often do well when given specific tasks. They can also organize well using time and deadlines. They tend to make great managers.

Think about how schools work: do this for forty minutes, then go to this class and do something else; manage your time well so that assignments in different classes get done at the correct time. Schools are really set up for left-brained people. Those who don't naturally work that way don't get work done within the given time period and end up getting in trouble or missing recess.

In addition, the people who decide to become teachers tend to excel in these situations. They are comfortable in the way schools are set up, so they become teachers and perpetuate the system. Most students seem to do fine within the system, and until class sizes decrease to the point that teachers have the time to get to know the needs of individual students, it's the best we've got.

Ask a left-brained person to improvise a solo in band, however, and you'll likely see the deer-in-the-headlights look. I remember asking the pianist in the high school jazz band how she knew which notes to play. There were no notes on the pages. She looked at me as if not understanding the question, saying, "I just play the chords." I had not had chord theory yet, so her answer did not enlighten me at all. I played notes on the page. Seeing the notation for a C major seventh—"Cmaj7"—meant nothing to me at the time.

I am more right-brained than left, but I'm one of those lucky people who can turn on the left hemisphere when needed. In school, I used the outlines and timelines and succeeded academically. I always knew, however, that my brain would prefer something else.

5—Nonverbal Ability

Nonverbal ability is primarily associated with the right side of the brain. I've discussed it a lot already in the sections about spatial perception and visual processing, but I wanted to give it its own chapter because verbal ability has its own chapter.

Our brains typically locate language in the dominant side of the brain, which, for most of the population, is in the left hemisphere. Verbal ability is the best single predictor of success in school. The question is: what happens when the individual is not language-dominant?

Right-brained Children

As I mentioned before, generalizations are risky, but sometimes we need make them anyway. Right-brained people don't tend to process information linearly. They don't work in step-by-step fashion. They excel at a global perspective. While so many people can't see the forest for the trees, the right-brained students see the forest every time. They see how it is integrated with the river and how the river is integrated with the ocean. Many can see how the whole is better than the sum of its parts.

I wish I had a quarter for every time a student said to me, "Just tell me what I need to know for the test." Few things would frustrate me as easily. I love to learn, but the beauty of learning didn't click for me until I figured out how concepts were linked when I was in college. For example, the Classical Period of music (approximately 1750-1820) included composers such as Mozart and Haydn. They composed amazing pieces of music using a well-defined format-: creativity within conformity. In comparison, the Romantic Period (approximately 1820-1870, including composers such as Beethoven, Wagner and Brahms) demonstrated much more individuality, much more focus on individual expression than rules.

What was happening during the Classical Period that might have inspired more individualism? The American and French revolutions, the establishment of the first democracy. The Classical Period was linked to more rules. But during that time period, the importance of the individual bloomed. And rules gave way to the individual expression prevalent in the Romantic Period.

In the early 1900s, Stravinsky wrote works for smaller ensembles rather than the huge orchestras used for romantic and impressionistic compositions. Before World War I began, he perceived undercurrents of conflict and believed that in time there would not be enough people left to still have large orchestras because people would kill each other off.

Right-brained individuals are often perceptive, empathetic, and creative. When they get into a project they enjoy, they lose all track of time. A professor once told me that the definition of ecstasy was the loss of the passage of time. By that definition, I have experienced ecstasy many times while composing music. It was not unusual for me to compose until I lost focus due to hunger. A musician who was in college at the same time I was actually had a doctor get involved and tell him he had to drink protein shakes because he would practice for hours and forget to eat.

Right-brained individuals are usually the ones who give us amazing works of art including paintings, sculptures, music, poetry, prose, or something else.

Humans need the arts. They are such a basic part of us that there is evidence left from primitive cave-dwelling times. They are how we express our humanity. Unfortunately, they are not the focus of most schools.

Time Is Linear

Take your memory back to the school schedule: forty minutes here, thirty minutes here, going from one subject to another. "No, you don't have time to finish that now, we have to go to gym class."

Many right-brained students are completely off balance when they are in school. And what do children do when they are off balance? They tend to either draw into themselves ("I don't fit here") or they act out ("I'm really not happy").

These children can drive their parents, as well as their teachers, crazy:

"Mikky, we need to leave in fifteen minutes!"
"Okay!"

And fifteen minutes later, Mikky is still doing what she or he was doing before Mom said anything. These children get labeled as oppositional because they don't have a concept of time.

These students are also ones who don't tend to get their work done on time. They spend far too many recess breaks staying in to finish homework. If anyone needs the break, it's the right-brained kid. They spend their days in a foreign environment—foreign to their brains, anyway.

As I said, I was lucky to be able to switch to left-brained mode when I had to for school. I was not happy though. College was better, without schedules that were locked in. There was so much more free time! I could spend more time on things I liked. Oops! That found me scrambling at the last minute to get the required work done. I would create schedules, doing biology homework here, math homework there. I wasn't very good at sticking to them, though.

I continue to be amazed at people who are organized. It's like the people who can make things look pretty, balanced, matched. I'm more like the absent-minded professor. Even though I got good grades through school, and as a school psychologist, I meet the time requirements of special education law, I am happiest when I don't have any appointments in a given day. I work on what I want to work on and switch topics when I want.

Step by Step

In math, right-brained students are often the ones who get the answer without showing their work. They do not do well if showing their work is required, because their brains don't think in steps. How can they show steps they didn't use?

Their thought process can be stymied by the teacher teaching in a step-by-step method, as I mentioned in chapter one. It's rather like the teacher saying, "Put this tree next to that tree, then this tree," etc. The right brained child is frustrated and thinking, "Why? Why? Why?" Until right-brained children know what the plan is or why they are doing a particular project, they often cannot take that first step. They can't get out of the gates, and they miss the entire lesson.

If they're really lucky, they will realize, "Oh! We are building a forest!" Too many are not that lucky, and will go home and tell Mom, "We worked on planting a bunch of trees together today, but I have no idea why. It was boring."

The teacher could help them a lot by just saying, "Today we will learn how to build a forest," at the outset of the lesson.

* * *

The problem with going to graduate school after teaching for many years is that I've had the opportunity to chide myself for all the things I could have done better. I had a friend in school whom I helped with math off and on from sixth grade through our senior year. I could quiz her the night before a test. She would have it. She could do the problems. Then she failed the test…over and over. In junior high, I went to the teacher after we took a test and vouched for the girl's abilities she had the night before. I know I was young, but if I had known anything about test anxiety, I could have addressed that rather than the math itself.

When I was teaching math or science, homework would count for half of the grade. Students who could not perform well on the tests still had ample opportunity to earn points by turning in their home-work. Showing work counted on the test, so I could learn their thought processes and know how to help them, along with giving them partial credit. Woe for the student who saw connections and was able to find answers without showing work, or, worse yet, turn-ing in homework. Right-brained students probably did not thrive in my classes.

Options

If you want different perspectives, ask a right-brained child.

One story I tell (all too frequently, I'm sure) about one of my daughters is that when she was about five years old, she put together some cardboard binoculars that were in a box of cereal. When I came downstairs, I praised her for her work, and picked them up to inspect more closely.

"Don't look through them!" she shrieked.

"Why not?"

"The box says they only magnify four times, and if you look through them, I'll only have three left!"

I could regale you with more stories of my daughter, but it would be easier for you to read the *Amelia Bedelia* books by Peggy Parish. They are hi-larious, unless you think about what it would be like to *be* Amelia Bedelia, always making mistakes and not knowing why.

Right-brained students often struggle with multiple choice tests. They're the easiest kind of tests, right? But when you can make what you perceive to be a logical argument for three of the four answers, they are not so easy. Knowing the material isn't the issue. Demonstrating that knowledge was the issue. All my daughter could do was try to think about what she thought her teacher might want her to think. She did much better on essay tests when she could just explain what she was thinking.

Essay tests are arduous to grade, however, so important tests like the ACT or SAT are multiple choice. I believe the SAT added an essay section in the late 1990s, but how many children struggled with them before that? How many were judged by themselves or others to be stupid because their brains didn't work well for multiple choice tests? I shudder to think, and I can only hope they found a way to value their brains in other ways.

Math

There are many right-brained individuals who do well at math. They not only can do the calculations, they can grasp the fundamental concepts and beyond. They include geniuses and inventors such as Einstein and Edison.

Just to demonstrate how complex the brain is

and how difficult it is to generalize, even though much of math is focused in the right hemisphere of the brain and spatial perception (including reading body language) is focused in the right hemisphere, many of the individuals who excel in math do not necessarily excel in spatial perception or reading body language. In fact, many folks who excel in math are confused by social structure and nonverbal communication.

The most profound examples of this are individuals with what used to be called Asperger's syndrome. It used to be included in the Autism Spectrum Disorder umbrella, but when the Fifth Edition of the Diagnostic and Statistical Manual of Mental Disorders (DSM-5) came out, most individuals who had been diagnosed with Asperger's syndrome were more accurately described under the diagnosis of Social (Pragmatic) Communication Disorder. The ones who continued under the Autism umbrella also demonstrated the repetitive movements and intense, singular interests.

These individuals do not understand or utilize language as greetings, sharing information, or as a part of the social structure. They do not naturally modulate their voices to fit different situations such as playground or library. They also do not have a knack for taking turns in conversation or expressing interest in the other person's needs.

For some unknown reason, there seems to be a link between excellence in math and lack of understanding of social communication. This was pointed out a few years ago when one researcher collected data demonstrating that there was a higher incidence of autism spectrum disorders in geographical areas such as Silicon Valley where mathematicians married mathematicians. I confess, however, that I did not read that study, but heard it on the radio. If you're not sure what I mean by the association between excellence in math and struggles with social situations, think of geeks. They frequently struggle socially. The positive thing is that these people tend to make a very good living; math it turns out, is one of the subjects prized in industries such as engineering and computer science.

6—Energy Levels & Emotions

Every person has different energy levels on different days as we slog our way through the work week. In addition to that, however, we tend to have a basic energy level that is more consistent over time. It's rather like weather, which varies more, compared to climate, which is more consistent in an area over time.

Introverts and Extroverts

If you have ever taken the Myers-Briggs test, you have learned about introverts and extroverts. There is extensive work available about the different personalities they describe. I want to take one aspect of what they describe and apply it to children.

Myers and Briggs point out that it is not that extroverts like people and introverts don't. It's much more complex than that. It's more about energy levels. Being around people invigorates extroverts. They get excited. Being around people drains introverts. Myers and Briggs point out that extroverts are excited about going to parties. They are the last to leave and go home pumped.

Introverts dread an upcoming party and often look for viable reasons to not attend. If they do attend, they are the first to leave. Their excuses, such as "I'm tired" or "I have a headache" often seem false. However, they *do* get tired or get headaches. Their energy levels have been zapped.

I read an interesting article years ago. The researchers did brain scans on people at rest. The resting brain wave levels of extroverts were much lower than those of introverts. No, you say, that's backward. But that was not correct.

An extrovert's resting brain wave levels are lower. Therefore, they are able to absorb a lot more energy from the environment without being overwhelmed. Introverts already have brain waves that are operating at a level with more energy. Any more energy they take in from their environment *can* easily overwhelm them.

Think about children coming in from recess. Most of them are still excited and loud. In fact, calming them down is often difficult for teachers, so many of the schools I've worked in over the years go from recess to lunch. There are some children, however, who come in exhausted, and it's not just from the physical activity.

Susan Cain, who wrote the book *Quiet: The Power of Introverts in a World that Can't Stop Talking* stated that introverts are not actually a minority as Myers and Briggs would have us believe. She may well be correct. It just doesn't match with my experience. Notice that I said experience, not research. Perhaps introverts devise coping methods in order to experience success in our intense world.

That said, think of all the people who go to scary movies or go on wild rides at amusement parks. They are looking for stimulation. Thousands go to sports events, political rallies, and parties. For me, going to Mardi Gras in New Orleans was one of the scariest events of my life. My husband and I went once when we were first married and never went back again. The crowd was so thick, it pulled me away from my husband and pushed me in the other direction. Some people would laugh. I was terrified.

Some people love white-water rafting or kayaking. They are most likely extroverts. I love canoeing. And when we go over a five-inch drop on the river, I squeal. Even my own family laughs at me, which is a bit funny because at least one of my daughters is as much of an introvert as I am. I watch sports events on TV to avoid the crowds. There was a huge march for women's rights after the recent election. I wanted to support them but could only do so in spirit and prayer.

It is also true that we can change over time. I am less of an introvert now than I was when I married my husband, who is definitely an extrovert. He, however, is more introverted than he used to be.

I remember once when I was in a musical as an adult. First, you need to understand that actors are not supposed to talk while backstage, as the sound can reach the audience. So we were back there, chatting *quietly,* when I mentioned that I was an introvert. Some other actors burst out laughing, and the director quickly came over and shushed us. Onstage, I do not present as an introvert. I admit it freely. But the key is that on stage I am playing a part. I am not me. I am comfortable singing in front of hundreds or even thousands of people, and I have done it many times. When I was in a living room and my parents would ask me to play for their friends, however, I was horribly nervous. I was me. The only way I could get through was to pretend I was by myself or playing a part. It makes sense, then, that other introverts would develop ways of coping as they grow up. They don't typically do it as children, though, and that's what is relevant here.

Schools

Schools are made for extroverts. To save money, districts increase class sizes and put as many children in one building as possible. My high school wasn't that large, decades ago. I think there were 800 to1,000 students. But there was one building. There are colleges with a thousand students that have campuses with many buildings. Of course, the dorms weren't needed in high school, but there are buildings for music, buildings for science, etc. In high school we were all locked in the building together for the day. Period.

These days, high schools often have several thousand students. They are still locked in one building, even though that building is huge. I had one patient whose anxiety was so high he had to be hospitalized. We eventually worked out that he would do online courses for half the day and attend school the other half. He was also allowed to leave class three minutes before the other students to avoid the overly crowded halls. The next year, he was old enough to take college credits that would count both toward high school and college, so that was even better. College campuses are less anxiety-producing than high schools. In high schools, every fifty minutes or so all the hundreds (if not thousands) of students stream from the classrooms into the hallways at the same time. If a student is not comfortable with crowds, this can be terrifying.

I have worked at an online school for seven years now. We have a lot of students who have too much anxiety about being in crowds for them to go to traditional brick-and-mortar schools. The online setting works well for some of them. The trouble we run into is that students who are successful in an online setting need to be very self-motivated. We do what we can, but a teacher who tells them to finish their work face-to-face often gets better results. That, however, is another topic.

A few years ago, a mother brought me a kindergartener who was a sweet boy who was becoming more and more aggressive at school. He hit other students and was getting in trouble a lot. As I chatted with him, he talked about enjoying coloring and art. When asked about joining sports teams, he said he didn't like them. I then asked him if he liked quiet activities or loud activities better, and he immediately said quiet ones.

I talked to the mother more and found out that the boy did most of his hitting when he was standing in line or at the end of the day. We tried an experiment. We arranged for the child to always be at the end of the line, unless it was his week to be the line leader. He was also allowed to go to the nurse's room a few times a day just to have some quiet time to "decompress." He stopped hitting.

In elementary school, classes eat together, learn together, play together, and go to the bathroom together. There is often no built-in time for quiet, and solitary time is simply out of the question. Children who are introverts by nature compensate in a variety of ways. Sometimes they pull into themselves and can develop symptoms of depression, especially when it seems that all the other students seem so much happier to be there. They can surmise that there is something wrong with them, which can reinforce depressive symptoms. Sometimes, they lash out, as with the boy mentioned above because they feel frustrated or cornered. Sometimes they develop anxiety about going to school and eventually about going anywhere.

It occurs to me that the old-fashioned methods of a strictly quiet classroom may have helped the students who were introverts. It was not the optimum way to help most of the children learn, but the children who were introverts were probably happier than in current settings where there are learning stations and group discussions.

Even as a grown woman, if I am asked to do some work at a high school, I try to time my entrance when the halls will not be filled with scrambling, loud students. One of the best things about attending college for me was that classes occurred at a variety of times, and the passing was done between buildings, not in confined hallways.

The Emotional Link

The difficulties that different energy levels cause us in school or at work naturally lead to a discussion of energy levels and their emotional links. Until now, I have described primarily aspects of learning and brain function. Our brains also directly influence our emotions, which, in turn, influence our brains. Both can have a tremendous effect on our ability to succeed.

For too many years, individuals who struggle with emotional issues have been labeled as weak, and worse. The light is beginning to dawn on society as more and more people realize that, again, our bodies work with electricity and chemistry. Emotions are chemical and electrical as well. And as with everything in our bodies, we have some, but not total control.

Years ago, I read the book *Molecules of Emotions* by Candace Pert, PhD. It blew my mind. I had never thought about it before, but it only makes sense that our emotions are not magically different from the way the rest of our bodies work. Our brains affect our bodies. Our bodies affect our brains. And the way they communicate is through chemistry and electricity.

Our understanding as a society is improving,

but there are still people who do not seek treatment for emotional or mental disorders because they don't want their bosses, coworkers, or lawyers to know they have a weakness. The sad thing is, they are probably right to do so. We all have weaknesses. No one is immune. What we seem to continue to struggle with, however, is that even though we all have different capacities, no one person has more value than any other one person. We may be in the minority, I fear, but I'm one of those scientists with a strong faith. I believe that God made each one of us in His image, and He loves all of us. I may have a lot of letters after my name, but God doesn't love me any more than my garbage man (for whom I am very thankful!), a neurosurgeon, or a homeless person living in her car. Sad to say, there are humans that perceive themselves to be more important than others. They will try to capitalize on the weaknesses of others to try to get ahead. They can't see that they are only able to get ahead on the material planes, not the spiritual plane, but that is a different topic as well.

Frankly, I don't even like the term "mental illness." There's only one body. We don't have regular illness and a separate category for feet. The brain is part of the body. I believe that, historically, "mental illness" was separated because of the stigma attached to symptoms believed to be associated with weakness. I would like to see the term stop being used. Depression is a health issue just as much as is indigestion.

When someone comes to me because they've been feeling depressed, the most common way to facilitate their healing is to help them see how important they are. What they do matters. Who they are matters. Everyone goes through times of feeling low and many experience times of hopelessness. With help, most of them improve. Why should there be a stigma attached to that? The important point in this book, however, is the effect that emotional difficulties have on children. Yes, children can suffer from depression or anxiety, just as adults can.

As a parent, the most important thing you can do is to make sure your child feels loved and important. No, they won't be able to do everything well. No one can. But they will be able to do something well. They have an important function within the big social picture, and you will love them no matter what. They may not be the football quarterback or the valedictorian of the class, but *no one* is better at being *them* than they are.

The Blame Game

Children are developmentally self-centered. Their brains have not developed to the point that they are able to see the big social picture. Remember the story about facial expressions and how they were interpreted? Children interpret everything through their own reality. Therefore, when their mother is angry, she is angry at them. When parents fight, it's because of something the child has done. That is one reason so many children have such a difficult time with divorce. Their parents will tell them that it isn't their fault, but they often don't believe it.

It might seem that that is not true because children are quick to blame others for something they may well have done. "It was Shawn's fault!" someone might say, pointing fingers anywhere but toward themselves. However, there are two things at play here. One, just because they may be fully aware that they are at fault, it doesn't mean they are mature enough to be willing to accept blame for it. Think of all the adults who will not own up to something they did. Children are even less inclined. They may blame themselves on the inside, but they don't want someone outside themselves blaming them. This survival tactic is so strong, many children will make up imaginary friends to take the blame for them.

The other thing that is at play is that children do not necessarily understand that lying about something makes it worse. From early on, parents need to be clear that if children tell the truth, they will not be punished as harshly as they would for lying about what happened. And they need to follow through with that. If this natural tendency for self-preservation is not curbed at a young age, it becomes associated with behavioral diagnoses such as Oppositional Defiant Disorder or even Conduct Disorder. From there, it is difficult to get the child back onto a path that will avoid criminal behavior.

Children need to be taught that telling the truth *is* a link to self-preservation. Every adult understands—hopefully—that everyone makes mistakes, everyone does things they shouldn't on impulse. When a child makes a mistake, then, those adults are typically willing to reduce the punishment if the child apologizes and is genuinely sorry.

Lying about the event, however, makes it more of a conscious act. Engaging the brain in a negative act is worse than not being able to control an impulse. Of course, we want to teach the child how to have better control over their impulses, but that comes with time. If children are not stopped from lying, that, too, will become ingrained over time. Since it is a conscious act, it will be ingrained more easily, and it becomes a part of the child's character.

As grim as the long-term effects of lying may be, it is important to go back and recognize that lying usually begins as a way to avoid punishment. If negative consequences are reduced for telling the truth, lying will be less and less associated with self-preservation.

It is shocking to admit that many adults appear comfortable with lying to avoid negative consequences, even if that negative consequence is only a perception. I do a lot of work with special education evaluations. There are forms that need be signed and parents may think that the school staff are just including the forms to make the parents' lives miserable, but they truly are legal requirements. The amount of paperwork can be frustrating for the school staff as well. Nevertheless, there have been many times that I will call a parent and ask them to please get the documents back to me. Their first response is to say, "I never got them." I know that things get lost in the mail. But I sent the form electronically, and I sent the hard copy twice. It is extremely unlikely that the parent never received the forms. If they just said, "Oh! I forgot! I'll get them out today," that would be the end of it. But they often go on to say, "You never sent them."

There are no negative consequences. No one is going to lock them up or charge a fine for not sending special education forms back in a timely manner. Somehow, however, they are so uncomfortable with telling the truth, that they feel the need to blame me for not sending the forms. I never know if I should point out that records are kept for both the electronic forms and the hard copies, but I typically just say, "How would it be best to get the forms to you?" I shake my head and worry about what they are teaching their children, but I admit that my temper rises a bit at the accusation. I am human, and I do not enjoy being blamed unjustly for something.

In order for parents to teach their children not to lie, they have to teach by example. That's tough. None of us are perfect at it, but we need to be aware enough to try. I love the Baha'i quote: "Truthfulness is the foundation of all human virtues." It's a good approach to live by.

7—The Power of Words

There is an old saying that "Sticks and stones may break my bones, but words will never hurt me." It's a big lie. A person can recover from a broken bone in six weeks. Hurtful words can haunt a person for the rest of their lives. Even though a person may tell herself or himself "That's not true," a little part of them believes it. And for children, it's a big part.

The Physical Reaction

Imagine someone saying something hurtful to you. "You are so ugly!" or "You are fat!" or "You are stupid!" Your heart rate increases slightly. Your cheeks warm. Your palms may even get a bit damp. Your stress hormone, cortisol, increases. There was no punch, but you feel like you have been hit in the gut.

Now imagine your significant other saying, "I love you!" or a friend saying, "You did an amazing job on that project!" That smile and warm glow are just the opposite of the stress attack of the cortisol. Instead, you get a little burst of dopamine, which makes you feel good.

My husband likes to tell the story of the man watching a shaman saying gentle words over a sick man. As they are leaving the tent, the man tells the shaman that he is ridiculous for thinking that gentle words could ever help someone who's sick. The shaman turns to him and yells, "You know nothing about it! You know nothing of our ways! You know nothing!" Then the shaman lowers his voice and says, "Do you feel the sickness in your belly? My words did that to you. My words can also make you feel better."

For decades, researchers have factored in the placebo effect when they test a new drug or procedure. They have to ascertain if whatever they are testing has more effect than the person has on herself. For example, if one group gets the new drug and the other group gets sugar pills, the group getting the drug has to exhibit significantly better results than the group with sugar pills or it's not the new product causing the change.

The improvement the group getting the sugar

pills gets is attributed to the placebo effect. It is perceived improvement based on the belief that the improvement should be happening. While some people laugh at a person's abilities to believe herself well, the point is, the improvement is there. Before we dismiss it, we should look more closely at what is going on. The patient believes they should be getting better, and they do. That implies that a certain amount of healing can be done with words and attitude.

In April 2012, the Harvard Health Letter wrote:

> For a long time, the placebo effect was held in low regard. If people responded to a suspect treatment, we said it was "just the placebo effect." The suggestion was that they had been fooled in some way, and their response was inauthentic. But attitudes are shifting, even in conventional medical circles. Randomized trials, some of them led by researchers at the Harvard Program in Placebo Studies and the Therapeutic Encounter, have deepened the understanding of the placebo effect and its various components. Researchers have also used brain scans and other technologies to show that there may be a physiological explanation for the placebo effect in many cases. There is some danger that uncritical acceptance of the placebo effect could be used

to justify useless treatments. But more important is the growing recognition that what we call the placebo effect may involve changes in brain chemistry—and that the placebo effect may be an integral part of good medical care and an ally that should be embraced by doctors and patients alike.

The placebo effect is controversial—not that it occurs, but if it's ethical to call it a treatment. When applied as an attitude rather than as a drug, and it has an effect, however, why not utilize it? It is a measure of a person's ability to influence their own health, or, in school, their own performance.

Now let's apply that idea to the children. If children are raised in a positive environment in which the adults support the children and their efforts, the children are healthier and have more belief in their own abilities. When raised in a critical environment, children do not develop confidence, and they also tend to cut others down to make themselves feel less inferior. In other words, they tend to bully other children.

Information about the importance of a child's self-esteem surfaced decades ago. Unfortunately, the information often translated into adults saying positive things about a child when they weren't true. If a child usually gets Bs in class and suddenly gets a

D on a test, the teacher has little basis for saying, "Good job!" In that case, I would pull the child aside and make sure everything was all right. Was the child physically okay? Were there some emotional issues going on at school or home? Did the student simply not understand the material that was tested, in which remediation might help?

Rather than lie about a "Good job," it would be better for adults to notice when a child exceeds expectations or even performs well in general. For example, if the student usually gets Ds and then earns a C, there is a legitimate reason to say, "Wow! That's your best grade so far!" A "C" might not be a good grade for Jenny, but may be a landmark event for Suzy and should be acknowledged as such. When a child is recognized for doing something positive, they are more likely to do it again.

The placebo effect can also work in a negative way, however. A teacher, coach, or parent that believes children will succeed typically gets better results than the adults who believe their children cannot do a particular task. Unfortunately, there are a lot of parents and teachers that do not believe in some students. Research has demonstrated that if you believe children will do poorly, they will. Expanding that idea, if you believe children will behave poorly, they will.

The movie *Stand and Deliver* was about a

teacher who believed his students could pass an AP Calculus exam. There was a book years ago about a teacher working in a poor area of New Jersey. It is called *Diary of an Inner City Teacher* by Tamam Tracy Moncur. There are others as well. The teachers would make a huge impact on both students and parents by calling home with *good* news instead of bad. The families had had such a long history of getting bad phone calls, they weren't sure what to do with good ones. How sad is that?

I work in the schools a lot, and I tell teachers and parents that 98 percent of students will behave well if they can. If they are not behaving well, it's up to us to figure out why. Bad behavior is frequently a reflection of not feeling important, not feeling loved, not succeeding academically, or not having friends. We have the power to improve all of those things.

Paradigm Shift

Society in the United States can be very judgmental. We grow up learning to criticize those who are different from ourselves or who think differently. This is not healthy for the general health of society, but it's also not healthy for the person doing the criticizing. Either they turn their critical opinions on themselves and are very unhappy, like average women comparing themselves to fashion models or they believe themselves better than those around them who do not agree with them. This can lead to narcissism and an unrealistic perception of their own worth and power. This never ends well. It might succeed for a while, but it never ends well. Think of Hitler.

I am not a fan of Freud. He had the attitude that if another person didn't agree with him, they were just too stupid to understand what he was saying. I believe it's more helpful for humanity if people don't automatically dismiss the ideas of those who have different beliefs, or think those who have different beliefs are stupid.

When I was in graduate school, I taught a course on educational psychology for students who intended to become teachers. We discussed how important it was for teachers to think well of their students. These college students were quick to state that they would approach all their students with a positive attitude.

I told them that was wonderful, but they needed to start that very day. I challenged them to have a positive thought about every person they saw as they crossed campus. They had to stop thoughts such as, "He's too fat," "Her skin is terrible," "That outfit looks terrible on her," "He looks greasy." Then they had to actively replace them with positive thoughts. I wish I had a picture of their faces!

They accepted the challenge, however, and we worked on changing the paradigm of their perceptions.

Unfortunately, critical attitudes are fostered early in the lives of children. I know a parent who thought it was funny that her four-year-old daughter criticized a grown woman, saying, "You can't wear black shoes with navy pants!" Personally, I don't believe any four-year-old should be criticizing adults, but the fact that the mother thought it was cute and funny ensured that the daughter would continue criticizing.

It's fine that the fashion world wants to make pretty clothes for people to buy and wear. However, to promote the buying of this year's styles, they also promote disparaging thoughts toward anyone not wearing the current styles. People who are smart think less of people who are not as smart. People who are rich think less of people who are not, as if a person's own value was based upon their bank account. These disparaging ideas are the basis of prejudice. They are also linked to a myriad of society's mental health issues.

For parents and teachers to help children grow up with positive attitudes, they need to demonstrate positive attitudes. Children raised with positive attitudes are more likely to have healthy self-esteem, not engage in bullying, not suffer with depression, and not consider suicide.

I apologize if I blindsided you with that last comment, but parents need to know that according to the CDC, suicide is the second leading cause of death for youth ages fifteen to twenty-four (behind accidents), and the third leading cause of death for children aged ten to fourteen (behind accidents and diseases such as cancer).

To pull it all together, positive attitudes can have a physical effect on the health of the brain. They can engage whatever mysterious control humans have over their own health, which has been labeled as a placebo effect. They help the health of the individual, but also of society, as they will eat away at prejudices rooted in faulty perceptions.

To be fair then, I suppose I need to say that Freud did at least get people to think about why they did certain things. I confess I can't think of anything positive about Hitler though.

Points to Ponder

Researchers at Harvard have rocked the medical world for years by demonstrating that patients with a strong faith heal more quickly and need less medication than those who do not report being persons of faith. In addition, those patients who have others saying prayers for them also heal faster with less medication. I don't really want to take a lot of time in this book about the power of prayer, but I want to point out its relationship with positive thinking in a scientifically researched application.

Prayer might be the ultimate positive attitude.

* * *

I certainly value the Constitution of the United States and the amendments associated with it. My question is if the first amendment is interpreted to support both positive and negative speech, should one child have the freedom to call another child names and bully them to the point that the victim takes their own life? That seems wrong, but I honestly don't know how to fix it. Perhaps it would help if parents and teachers raise children to understand that their own negative thoughts and opinions have negative impacts on their own health in addition to the person they are insulting.

8—When Things Go Wrong

With one hundred billion cells in the brain, it is highly unlikely that all of them are arranged in exactly the correct place and functioning at optimum capacity at any given time. I suspect that few, if any, people have everything in their brains working perfectly. We all fall down. We all imbibe something that has a potentially negative effect on our brains, such as caffeine, alcohol, or drugs (this includes over-the-counter and prescription drugs). It's a miracle that we get along in life as well as we do.

Diagnosing Children

There are several diagnoses associated with children. In addition to them experiencing anxiety and depression, as already mentioned, pediatric psychiatric diagnoses include: Traumatic Brain Injury (TBI), Attention Deficit Hyperactivity Disorder (ADHD), Autism Spectrum Disorder (ASD), Communication Disorders, Intellectual Disabilities, Specific Learning Disorders, Motor Disorders, Oppositional Defiant Disorder, and Conduct Disorder.

Communication Disorders typically fall under the purview of Speech and Language Pathologists, so only one will be mentioned here: Social (Pragmatic) Communication Disorder. In addition, Motor Disorders that relate to my practice are generally tic disorders or Tourette's. Although motor disorders are rare, they are extremely frustrating for the person experiencing them. Treatment for the actual motor problem often falls to a neurologist, but they are included in the psychiatric diagnostic manual because the diagnosis can be associated with poor self-esteem, anxiety (especially about being in public), and impaired function in daily life.

TBI

Additional research is coming out every day about concussions. Years ago, it was believed that a person had to lose consciousness to be diagnosed with a concussion. Now we know better. The CDC defines TBI as, "a disruption in the normal function of the brain that can be caused by a bump, blow, or jolt to the head, or penetrating head injury. Everyone is at risk for a TBI, especially children and older adults."

For example, in a car accident you don't have to actually hit your head to sustain damage to your brain. The violent shaking of the brain caused by the jerking of the head can cause damage at the cellular level. The damage may or may not show up on an MRI or CT scan. The brain is about the consistency of day-old Jell-O. It doesn't take much to shake it around, especially since the bones of the skull supporting the brain are sharp in the front and on the sides.

In 2010, the rate of head injuries reported for emergency room visits including visits, hospitalizations, and deaths was 823.7 per 100,000. The current population of the USA is 323 million. If you do the math, that works out to over 2.6 million reported TBIs *per year.* Experts agree that at least that many more go unreported.

When your child hits his or her head, watch for symptoms of headaches, a change in memory, or actually any changes. Be especially careful if a lump is raised or if stitches are needed.

I always ask parents three times if their child has had a head injury, as taught by my neuropsychology professor. The first time, the answer is "no" at least 95 percent of the time. When I ask again, specifically about cuts and lumps, they begin to consider. I have actually had more than one parent state three times that their child has never had a head injury. The child is sitting right there. Then I say, "How did he get this scar here across his forehead?" *Oh yeah, that.* It's just that people don't note the smaller injuries. If their child didn't lose consciousness, they don't note the event.

When I was about four years old, I had a bad fall. I hit my head so hard, my baby teeth went up and knocked one of my permanent teeth crooked. I had several seizures and was treated by the neurology clinic at Iowa University Hospitals. I was put on phenobarbital for approximately six years. Phenobarbital is a barbiturate, meaning it slows down the central nervous system. I blame my slow pace of walking on the phenobarbital, which I'm sure has no foundation at all, but helps my hyperactive husband remember that I'm just not made fast.

I was lucky. The damage didn't seem to affect my cognitive ability, as I basically aced school. But don't ask me to walk a straight line. I don't drink, but I know I couldn't pass the balance tests if I did! I also can't sleep without medication, which has its own problems, but in the grand scheme of things, I consider myself lucky.

If your child has a head injury, watch for difficulty remembering things and short tempers. When a parent brings me a child who can have tantrums that go from zero to sixty in three seconds, TBI is the first thing I check for in the medical history. Here is a list of symptoms developed for a handout for one of my presentations:

- Memory deficits
- Reduced ability to plan and follow through
- Distractibility
- Impaired awareness of time
- Impulsiveness
- Lack of initiative
- Seizures
- Impairment of spatial perception
- Word retrieval problems
- Explosive temper outbursts (often with little provocation or escalation)
- Headaches

If you notice these symptoms in the days following the accident, it would be wise to try to get in to see a neuropsychologist. Of course, you'll probably need to see your family physician first. I also want to add that for children with a TBI, their tantrums will often last longer than those of more typical peers. They are often exhausted afterward and are frequently remorseful. It's rather like an electrical explosion in their brain that happens *to* them rather than just being angry for not getting their way.

The unfortunate thing about long-term effects of a mild TBI are that it is possible to damage a part of the brain, and the damage won't show up until the developmental stage that uses that part is reached.

For example, if a six-year-old child has a TBI that damages cells in the brain that will later develop to help him learn algebra, we won't know until he is in eighth grade that there's a problem.

I've just been sitting here watching America's Funniest Videos. I think there might be two million head injuries each year just on that show alone.

ADHD

There are a lot of ADHD diagnoses out there, and there are several reasons for it. First of all, ADHD, like other diagnoses in the DSM-5, is defined by a cluster of symptoms. There is not a single cause that can be documented. For example, there is a small lobe-like structure on each end of the corpus callosum, the bundle of nerve fibers that connect the two hemispheres of the brain. One study done in Georgia years ago pointed out that some individuals with ADHD had a smaller lobe or no lobe at all on the front of their corpus callosum. This inhibited the communication between the two halves of the frontal lobe. Since the frontal lobe is associated with staying focused and making good judgements, it makes sense that these individuals would meet criteria for ADHD. By the way, it's always called ADHD now. There is no ADD. The psychologist diagnoses ADHD, then specifies if it is the hyperactive/impulsive, inattentive, or combined type.

Another "cause" of ADHD is the slow development of the frontal lobe. The frontal lobe is the last part of the brain to develop, both in individuals and evolutionarily. If a child's brain does not develop at the same pace as his or her same-age peers, the child will not demonstrate age-appropriate behaviors of staying on task, etc. A diagnosis of ADHD might help those children succeed in school with accommodations. That is one reason why so many children with ADHD are described as acting younger than their age.

If you recall the earlier explanation about how brain development is impacted by environment, you can picture children in a dangerous environment needing to monitor their whereabouts at all times. "Dangerous" could mean homelessness, family violence, neighborhood violence, or other scenarios. If children have to constantly monitor their environments, it makes sense that their brains will keep connections needed for paying attention to everything and sacrifice those connections meant for focus on things one at a time like school work. When these children struggle in school, they are often diagnosed with ADHD.

This makes me think of what I read in a book about dog training. The book said it was important to keep the dog at heel during walks rather than letting them be several feet in front of their master. The author said that if the dog is at heel, the human becomes the alpha, with the responsibility of checking for dangers. If the dog is allowed to be in front, it will assume the responsibility of watching for danger. The author says that having a dog always walk in front is detrimental to its mental health! Apparently, that is one reason they go crazy and run in circles in the house. So having to monitor surroundings for danger might be bad for human's mental health as well.

I watch for significant differences between areas of cognitive functioning. If Glen has average verbal ability and nonverbal ability, but his processing speed score is significantly lower, I will consider ADHD. There are other tests I give, but significant differences are red flags for me. The good news is that there are a lot of different medications for ADHD. For the stimulant medications, it will usually be clear in two or three days if they are going to have a positive effect. For those people who don't tolerate the stimulant medications, there are other options. They take longer to build up in the body before an effect is clear though. I always tell my patients to have patience. It takes a while to find the right medication and the right dosage, but it will be worth it in the end.

Contrary to many popular beliefs, research has also shown that children and adolescents diagnosed with ADHD have a higher incidence of chemical/alcohol use when they do not receive medication than when they do. They tend to self-medicate.

There's a paragraph I put in every report in which I diagnose ADHD. It is from the DSM-IV (Diagnostic and Statistical Manual of Mental Disorders—fourth edition). Similar information is in the DSM-5, but it's not in one neat paragraph. ADHD in the DSM-IV is described as:

> Associated features vary depending on age and developmental stage and may include low frustration tolerance, temper outbursts, bossiness, stubbornness, excessive and frequent insistence that requests be met, mood lability, demoralization, dysphoria, rejection by peers, and poor self-esteem. Academic achievement is often markedly impaired and devalued, typically leading to conflict with the family and with school authorities. Inadequate self-application to tasks that required sustained effort is often interpreted by others as indicating laziness, a poor sense of responsibility, and oppositional behavior. Family relationships are often characterized by resentment and antagonism, especially because variability in the individual's symptomatic status often leads others to believe that all the troublesome behavior is willful. Family discord and negative

> parent-child interactions are often pres-
> ent. Such negative interactions often di-
> minish with successful treatment. (87-88)

The child may appear lazy, but it's difficult to initiate a task when you know it's going to take a lot of concentrated effort. It's like an adult deciding to postpone a complex task because it's eleven o'clock at night, and she will be fresher in the morning to face the task. Except that those with ADHD don't often get that fresher time when they can accomplish something without a lot of effort. They really are not behaving like this to make your life difficult, although I know there are times when it seems like it. Individuals with ADHD are often accused of being unmotivated, but I really feel that the majority of them are doing the best they can.

ASD

A child on the autism spectrum usually starts displaying symptoms as a toddler. These children often seem to be in their own worlds. They prefer to play by themselves, with no desire to interact with other children and no clue how to go about doing it anyway. Early intervention can have wonderful results.

These children also display repetitive motions such as rocking or hand flapping. In fact, repetitive patterns of behavior, interest, or activities are required to be present in order to make a diagnosis.

They do not deal well with change. In schools, these children can tantrum when presented with change to the point that they hit and bite their caretakers. I know of a school district in Minnesota that got in trouble with the media who told a story about caretakers and teachers in the autism program who put children in restrictive holds. Unfortunately, the reporter didn't bother to explain that these children could get quite violent, and it was not unusual for the school staff to go home with bruises, bites that broke the skin, and, in at least one case, a broken arm. The staff put the children in holds for everyone's safety.

That said, I know of another Minnesota school district that frequently uses restrictive holds as a first, rather than last, resort. It's never a clear picture. Generally, these days holds cannot be used to protect property—only people.

There is often a lot of confusion around autism. One reason is that the clinical criteria for a diagnosis of autism is not the same as the educational criteria. A child can meet criteria for educational autism and qualify for special education services, but they won't meet criteria for a clinical diagnosis. The opposite is true as well.

For the majority of these children, language develops late, and it can be impaired throughout life.

Communication Disorders

We don't call these CDs because that acronym is typically associated with Conduct Disorder.

The only communication disorder I really want to mention here is Social (Pragmatic) Communication Disorder. Individuals who qualify for this diagnosis typically do not understand how social relationships work. They don't understand the give and take of conversation, and nonverbal communication is usually like a foreign language to them. In the past, these individuals were included under the autism spectrum and the term "Asperger's syndrome" was used. Asperger's is not listed in the current diagnostic manual. I do know that a lot of folks were essentially diagnosing themselves with Asperger's if they were a bit awkward socially. In addition, several states have funds available to help support individuals with autism when they are unable to support themselves. There might have been a benefit to some by having Asperger's under the autism spectrum umbrella. Perhaps it was abused. I don't know.

Anyway, individuals with Social Communication Disorder may not struggle as much as those with moderate or severe autism, but they certainly struggle. These individuals are differentiated from individuals with autism because they lack restricted patterns of behavior, interests, or activities. In addition, language development usually occurs on time with individuals with Social Communication Disorder. It is late when autism is present.

Intellectual Disabilities

The DSM-5 was published in 2013 by the American Psychiatric Association. Believe it or not, the fifth edition was the first time the word "retarded" was not applied to these individuals. Of course, "retard" means to slow, and the individuals with this diagnosis are often described as slow, but the word has been used in such negative ways over the years, I am truly glad that they got rid of it.

At this point, it is crucial to point out that, again, every person has an important part to play. Cognitive tests such as the *Wechsler Intelligence Scale for Children-fifth edition* (WISC-V) or the *Woodcock-Johnson Tests of Cognitive Ability- fourth edition* (WJ-IV Cog) essentially measure an individual's ability to succeed in school. Those with scores of less than 70, or at the second percentile when compared to their peers usually don't have academic success in school. Their curricula focus on learning how to live as independently as possible. On the other hand, I have seen, repeatedly, a child with intellectual disabilities come into a family and teach the other members how to be patient and how to love. That is no small thing. I have not met one parent who has not blessed their child with intellectually disabilities for coming into their lives.

When I was in graduate school, my professor for child development told us about a study in Russia. Half of the children in an orphanage were treated the way they had always been treated. All of their needs were met. The other half, however, also had adults with intellectual disabilities nurture them. The adults were not responsible for their care, only to hold, comfort, and play with them. After several years, the children who had been given additional nurturing averaged twenty points higher on IQ tests. Twenty points is a lot. The average range spans twenty points and includes the middle 50 percent of the entire population. It was a remarkable effect caused by people with intellectual disabilities. An intellectual disability does not translate to low-worth.

Specific Learning Disorders

In schools, these are called Specific Learning Disabilities (SLD). The current DSM-5 is now closer to the definitions used in schools, but there are differences.

Clinical: children can be diagnosed with a disorder in mathematics (calculation or math reasoning), reading (phonics, fluency, or comprehension), or written language (spelling, grammar or organization).

In schools, there are eight areas of learning disabilities as defined by the federal government. There are the same mathematics and reading disabilities, but the area of written expression does not include spelling or grammar. The last two areas are listening comprehension and oral expression. These last two areas are not listed under the clinical disorders.

After World War I, injured men were found to exhibit symptoms we now associate with learning disabilities and TBI. A lot of testing, research, and descriptions of symptoms occurred. I find it fascinating, but I won't go into it all here. It would fill an entire book all by itself. Nevertheless, that was the start of research defining learning disabilities.

It wasn't until 1975, when the federal government passed a law stating that every child would be educated in the public schools, that children with disabilities started receiving a wide range of special education services beyond intellectual disabilities, blindness, and deafness. The wider range included students with learning disabilities.

There is still controversy connected with how to determine if a child has a learning disability. Currently, if there is a significant difference between a person's cognitive ability and their academic achievement, that person is said to have a learning disability. To make matters a bit more confusing, different states have different definitions of that significant difference, so a student could potentially qualify as having a learning disability in one state and not another.

In an effort to be more consistent, research at universities came up with a process called Response to Interventions (RTI). Teachers would use research-based interventions, and if those didn't prove effective after several weeks, the child would qualify as a student with a learning disability. Problems inherent in that system are that the fidelity of the interventions again falls on the general education teachers who are already overwhelmed by all of the current demands placed on them. Very few schools are able to incorporate well-documented interventions for the RTI method. Therefore, the majority of schools are continuing to use the discrepancy method.

In an effort to recognize that some children struggle in a classroom because they learn differently than the teacher's teaching methods, many individual school districts require that scientifically-based interventions be done before a student can qualify as having a learning disability. This is an attempt to combine the two approaches, but, unfortunately, the responsibility of interventions falls on the overwhelmed general education teacher.

Individuals with learning disabilities generally have average or above average basic intelligence, but there is a glitch somewhere in their brains that prevents them from performing up to their abilities in an academic area, even though they have received education in that area along with other subjects. For some unknown reason, one particular area just didn't click.

Along with that significant discrepancy, the child must demonstrate a difficulty in some area of processing information. For example, auditory memory might be at the same level as their overall abilities, but their visual memory is much lower. All types of memory might be on a par with overall abilities, but processing speed is low. In addition to support in their academic area of need, then, the child also gets support in the area of processing that is low.

A straightforward example would be if a student has average intelligence but demonstrates a weakness in processing speed. In academic areas, the child scores in the average range for math, written language, and even basic reading, but reading comprehension is significantly lower. That student would qualify for special education support for reading comprehension, but since that could well affect all subjects, including math story problems, the student would get help wherever it was needed. For example, he might need support in history, since there is so much reading involved. It often helps these students to have the material read to them, and these days, thousands of books are recorded for these students. There are also computer programs that can read texts.

Because of the slow processing speed, the student would likely qualify for supports such as class notes provided by the teacher prior to class, shorter assignments (because homework can take twice as long with a weakness in processing speed), and more time for tests. Recall that taking notes is a very complex task requiring adequate short-term memory and processing speed. If either one of those is lacking, the student will struggle. In addition, if you've ever tried to read another person's notes, you know that getting the notes from the teacher is much more accurate. The added advantage is that the student with the learning disability can follow along during class rather than just sitting there missing every other word.

Problems arise, however, because very few cases are so clear. For example, in the state of Minnesota, a student with an IQ of less than seventy-five cannot qualify for special education under SLD. This likely came from the early definition of SLD as a child having "average intelligence." The average range is 90 to 109, so even an IQ of 85 might not qualify at first blush. The people who wrote the law realized that 90 to 109 is much too restrictive. Nevertheless, a child with an IQ of 75 is likely to have a lot more difficulties than a possible learning disability.

Yet what about the student with an IQ of 74 and a significant difference between cognitive ability and academic achievement? She can't qualify, at least in Minnesota. I suspect every state has a cut off number though. In Minnesota, if a student has an IQ of 72-74, there is no special education support for them unless they demonstrate significant behaviors or have a diagnosis. The law recognizes error of measurement, so if a student has an IQ of 72, the school might be able to make a case for the presence of an intellectual disability. I remember years ago, however, a student who had an IQ of 73. There was a dramatic difference between her cognitive ability and her academic achievement scores, but she could not be said to have an intellectual disability (IQ too high) or a learning disability (IQ too low). She was sweet as can be with no behaviors to address, no behavioral disability, and no medical diagnosis that could qualify, so there was no Other Health Disability (OHD). Therefore, with a low IQ and academic achievement below the 0.1 percentile when compared to her peers, special education support could not be provided. Cases like this are probably the best argument for alternative methods of qualifying students.

Another complication is the child with a high IQ who also has a significant discrepancy between cognitive ability and academic achievement. If their achievement scores are in the average range even though they have a gifted IQ, they could meet the discrepancy criteria, and they could well meet the information processing criteria. Many schools will balk, however, at providing special education services to a student with average academic scores. In my experience, some schools will, and some won't, but the ones who will do so because the student is exhibiting behaviors related to their frustrations with their learning problems. I have reasoned with some people through an analogy: "Suppose you have a brand-new Corvette, and you're out on the highway. It gets stuck in a lower gear, so you can go at a speed like the other cars, but it is still frustrating." They usually get that.

Oppositional Defiant Disorder

There are days when any child on the planet could meet criteria for Oppositional Defiant Disorder. The DSM-5 lists the symptoms as:

- Often loses temper

- Is often touchy or easily annoyed

- Is often angry and resentful

- Often argues with authority figures or, for children and adolescents, with adults

- Often actively defies or refuses to comply with requests from authority figures or with rules

- Often deliberately annoys others

- Often blames others for his or her mistakes or misbehavior

- Has been spiteful or vindictive at least twice within the past six months

The caveats are the importance of the word "often," as well as that the child has to exhibit at least four symptoms from the list, and the behaviors have to have occurred for at least six months. That's what helps narrow down the field.

* * *

Knowing that most children can be this way at times, how does a psychologist determine if a child should be diagnosed with ODD? First of all, experience and training count for a lot. The key, though, is that the behaviors regularly interfere with the child's functioning. It is also important to rule out other causes.

For example, children with ADHD can also lose their tempers easily, as can children with depression. Irritability is a symptom of childhood depression.

Children with autism are often easily annoyed, although there are other symptoms present that make differentiation fairly easy.

Children whose parents are going through a divorce and those dealing with the addition of step-siblings can often be angry or resentful. It can also be present when a child or youth believes there is something wrong with them—perhaps they are dealing with a learning disability—and they perceive unfairness when their peers do not appear to struggle.

Arguing is fairly typical for individuals with ADHD. Some people enjoy it so much it's almost like a sport with them. Dr. Amen, author of *Change Your Brain, Change Your Life*, says that individuals with ADHD will push other people's buttons as a form of stimulation they need. That aside, however, some of the most argumentative children are the gifted ones. They know what is right, and they don't care if you are an adult or not. Even though they can be difficult, try to be a bit understanding. Life is very difficult for a gifted child. When my daughter was in second grade, she was reading at the sixth-grade level at least. Her teacher would not recommend her for the gifted program, however, because she was well-behaved! Thankfully, the principal stepped in.

The openly defiant kid is the student who owns his own world, including you. There is a lot of controversy about these kiddos. Are they born, or are they raised that way? Sometimes it's a lack of empathy for others, and sometimes their own lives are so difficult, there's no energy left for empathy. Sometimes their parents spoil them, and they feel entitled. Although it's typical for toddlers and preschoolers, these natural behaviors should be trained out by the time the children get to school. Ask any teacher. That's not happening with all students, although, honestly, it does with most.

A child who deliberately annoys someone can be a bully, but they can also be seeking attention. For many children, even negative attention is better than no attention. Teachers continue to work to learn to pay more attention to the children who are behaving well. It's so easy to be sucked into the child's misbehavior! A child who annoys others on purpose can also be the child with ADHD who aggravates others to stimulate his own brain.

Blaming others for one's own mistakes is quite natural. This was addressed slightly when talking about lessening punishment when the child tells the truth about what happened. The denial of culpability stems from a fear of what the punishment will be. What we're talking about in this case, however, is the child who blames someone else because they want the other child to suffer the consequences. They want the child to suffer. This is not just a lack of empathy but a delight in the suffering of others. When this occurs, psychologists do an environmental analysis and typically start treatment as early as possible. This ties into the vindictiveness mentioned later.

When considering a diagnosis of ODD, psychologists have a lot to consider. In general, I still believe the vast majority of children will behave well if they can. Very, very few take a true delight in the suffering of others.

Conduct Disorder

If a diagnosis of Conduct Disorder is being considered, the child, or more likely, youth, has probably
been involved with the law. These are the youths
who set fires, are cruel to animals, and who run away.
These cases are extremely complex, and I don't think
this book is the correct format to address them. These
children have often endured abuse. They are frequently taken out of the home, but that presents its
own problems. Most children, even when they have
been abused, will prefer to stay with their parents. It
takes a dedicated team to help these children.

9—Examples

At this point, I thought it would be informative to take you through a few cases to understand some of the thought processes involved. First of all, it might help to understand a few things. The tests used for measuring cognitive ability are standardized. That means that before they are published, they are administered to hundreds if not thousands of individuals. As would be expected, the majority of people score in the middle, with decreasing numbers as scores get higher and lower. The larger number of individuals in the middle is what generates the shape of the bell curve. I realize that the bell curve itself can be somewhat controversial, but that is another topic. At this point, let us just go forward with it being an acceptable way to compare scores.

For the majority of cognitive batteries, the average range is defined as the middle 50 percent of the people tested. There are a few exceptions, but this is not the format in which to address them. If the middle 50 percent are average, then, by definition, 25 percent of the population is above average, and 25 percent of the population is below average. So when politicians say they are going to boost children who score below average up into the average range, please know this is impossible. If children who are in the below average range score higher on subsequent testing, the entire chart merely moves to the right.

The only way for a child to move from the low average range to the average range at the next testing is to improve relative to his peers. Wait a minute. That sounds confusing. Let's try this. Imagine a symmetrical hill gradually tapering down into the flats below. Now, for everyone hundred children tested, fifty of them are clustered on the hill, with numbers of children gradually decreasing until there is only one child out on the plain on either end. In order to move to the right (improve his score), a child has to move past the other children.

It is possible for an individual to improve his or her score, but for that to happen, there has to be an opening further along. But for that to happen, the individual has to do well enough to move another person down lower. Of course, no child scores exactly the same on subsequent test administrations. That's why scores are always reported with a confidence interval. For example, if Terry earns a score of one hundred on a test for verbal ability, there would be a 95 percent chance that if tested in the future, Terry would score somewhere between 92 and 108. So, the children on the hill are constantly in motion, shifting back and forth. The range for Jacob might be 85 to 94, which places him in both the low average and average ranges.

Without going into a description of statistics, which I'm sure you'd rather not delve into deeply, let's move forward with the bell curve, standardized testing, and cognitive test batteries to provide information about how brains work.

When dealing with standard scores, it is important to know that a large difference between test scores weakens the accuracy of the cluster, or summary, score. An average of eighty might give you valid information about someone with scores of seventy-eight and eighty-two, but it doesn't give you valid information about someone with scores of sixty and one hundred. That person has an area of weakness and an area of strength.

A difference of fifteen points is notable; a difference of over twenty-two points is statistically significant. In addition, individuals with significant differences between various cognitive abilities tend to experience higher levels of frustration than their peers.

The majority of people have their scores for various aspects of cognitive ability within about ten points of each other. The majority of people that come in for a neuropsychological evaluation typically have differences of more than twenty, and I've seen fifty- or sixty-point differences several times. These people come for an evaluation because they are dealing with symptoms that are interfering with their daily functioning.

All of the cases below are conglomerates of cases I've had over the years. The names are fictitious. We'll define the average range as 90 to 109, the low average range as 80 to 89, and the high average range as 110 to119. This will cover over 70 percent of the population. To save space, the individual test scores have been eliminated—unless there is a significant difference between test scores present. As stated above, that weakens the summary score. Please know that these cases also involved other neuropsychological tests. I'm only taking you through the cognitive results.

Case 1

A child named Kinsey was referred by a pediatrician for possible ADHD. I administered a cognitive test battery called the *Woodcock-Johnson Tests of Cog-*

nitive Ability. Kinsey was very talkative, sometimes not waiting for others to contribute to the conversation. She impressed her third-grade teacher as being very smart, but she did not do well in school.

Looking at the test scores, Kinsey exhibited a personal strength in her verbal ability—utilizing and understanding language. She could express herself well. Her verbal ability was in the high average range, ranking her at the 87th percentile. That means she scored higher than 87 percent of her peers. Kinsey was also well-behaved, except that she did not get her school work finished on time.

Kinsey had average abilities for her age for nonverbal, a visual-spatial thinking score of 96, and a long-term retrieval score of 102. These scores indicated that she was able to learn new information, process visual information untimed, solve visual puzzles, and recall information longer than twenty seconds at a typical level for her age.

The two tests used to generate the short-term memory score differed by twenty points, weakening the summary score. One was in the average range, and one was in the high average range. At this point, there doesn't seem to be a problem.

Next, we come to Kinsey's score for processing speed. Her score of seventy-eight was thirty-nine points lower than her verbal ability. Recall from above that twenty-two points is statistically significant, and many students with significant differences between areas of cognitive ability experience higher levels of frustration than their typical peers. The fact that Kinsey was well behaved was commendable. She had two parents living in the home with her, and they were loving and supportive. That likely had a positive effect.

With a processing speed slower than 93 percent of her peers, Kinsey had weaknesses in both auditory and visual processing in real time. The testing that produced scores in the average range for both were done in a quiet, one-on-one setting, and they were not timed. In real life, conversation takes place in real time. The words are there, then they're gone.

As mentioned before, imagine that you're just going out the door, and a member of your family says something to you. You miss a chunk of it, but you don't want to go to the trouble to go back into the house to confirm what was said. Therefore, you replay what you did hear, and then piece together what must have been said. Sometimes you're right. Sometimes you're not.

When people have slow processing speed, it's like that most of the time. They seldom have time to replay conversations or teacher's instructions, because other information is still coming at them. Our brains don't like incomplete information. Sometimes they fill in gaps without us even being aware of it. Sometimes they're right, but often they're wrong.

Now Kinsey did not have any behaviors of concern, but with such a slow processing speed, she often missed directions. She also had difficulty completing her assignments in the time allotted. For example, say the teacher taught a math concept for twenty minutes, then gave the students fifteen minutes to complete six problems before it was time to go to music class. Because she missed the oral directions, it would likely take Kinsey at least five minutes to figure out, either by watching other students or by going up to ask the teacher, what she was supposed to be doing. After she figures it out, she's working, but less than half as fast as her typical peers. If they do six problems in fifteen minutes, she's only able to do two or three. Recall, however, that it took her ive minutes to figure out what to do. Therefore, she is only able to do part of the assignment before the teacher says, "All right, class. Put your math away. It's time to line up for music." Now she has homework, but most of the other students got their assignments done.

The numbers are arbitrary. What is important to realize is that with such a slow processing speed, there is little chance for Kinsey to be able to finish her math or any other assignments at school. At the end of the day, she tries to pack up the homework she didn't get done during the day. Few of the other children have any homework because they got theirs done in the time allowed by the teacher. This could cause her to feel sad, angry, or frustrated.

Even if there were only an hour's worth of homework that was done during the day, it will conceivably take Kinsey two to three hours to finish the same amount of work. The most frustrating part is that even if she gets the work home and doesn't mind having to do homework when other kids already have theirs done, when she sits down to do the work, it's likely that she can't remember what she's supposed to do.

After the results of the evaluation, Kinsey got support at school. She received help from the teacher and the paraprofessional in the room to organize her assignments at the end of the day. She took them home with written instructions, so her parents could help her with the work. Kinsey also was given shorter assignments so she wouldn't have to spend more hours a day doing assignments than her peers.

At home and school, Kinsey got visual reminders of tasks that needed to be done. If oral directions were necessary, she was only told one thing to do at a time. She learned to repeat the instruction before she started it, to confirm that she had gotten it right.

Every person needs to do something each day that they are good at. Kinsey loved to use her verbal ability to make up stories. She was given about fifteen minutes a day at school to make up stories. She made a recording so that she was not limited by how fast she could write. She definitely had no difficulty talking quickly! Her mother worked with the teacher to transcribe the stories and print them out. When Kinsey's homework was done, she was allowed to draw pictures to go with her stories. She loved it!

KINSEY	Standard Score	Classification
Verbal Ability	117	High Average
Visual-Spatial Thinking	96	Average
Processing Speed	78	Low
Short-term Memory	110 (weak score)	Average
Long-term Retrieval	102	Average

Case 2

Derek was in seventh grade, and he was receiving special education services under the category Emotional Behavioral Disability (EBD). He attended a school in which cognitive testing was not required for EBD students. An analysis of their behaviors and what the teachers perceived as driving those behaviors was done. Utilized in special education evaluations, it is called a Functional Behavioral Analysis (FBA).

The educational team determined that Derek wanted attention from his peers and was trying to avoid school work. They put programs in place to try to get him attention for positive behaviors as well as rewards for completing his homework. The team concluded he was capable of doing the work, but he was not a motivated student. His behaviors were not improving, so his mother brought him to me for an evaluation.

Derek's scores were in the very low range overall. There were not significant differences between areas of cognitive functioning. The school, however, was asking him to process and understand curricula that 99 percent of his peers found easier than Derek did.

There is one category of special education that has been renamed many times over the years. Names have included "retarded" and "cognitively impaired." The current terminology is Developmental Cognitive Disability or DCD. There are individuals whose abilities are just really low. To put them in classes made for students with higher abilities would not be fair to them. While some people find this a controversial topic, I believe it is important to consider fair as being what's fair for the child.

However, the real reason I included this example was because there are states that do not use cognitive testing, using the reasoning that it labels students. The school was already labeling Derek as a student who was not motivated, would not do his work, and acted out to get attention from his peers. Wouldn't it be fairer to Derek to recognize that he was not capable of doing the work being presented to him instead of blaming him for not doing it?

DEREK	Standard Score	Classification
Short-term Memory	68	Very Low
Visual Processing	75	Low
Long-term Retrieval	63	Very Low
Nonverbal Fluid Reasoning	65	Very Low
Crystallized Verbal Knowledge	71	Low
Fluid/Crystallized Intelligence (FCI)	64	Very low

Cases 3 & 4

Kaylah and Joshua exhibited significant differences between scores *within* cognitive areas rather than *between* the cognitive areas. Instead of true weaknesses, their weaknesses were being able to sustain attention long enough to get consistent scores.

Kaylah was referred to me because she was exhibiting signs of depression, and she was struggling academically. For Kaylah, her verbal ability was just into the high average range. The tests used to generate the summary score were close to each other, making the summary score reliable. Her verbal ability was her personal strength. That worked well for her because language is so important in nearly every subject. Her scores for nonverbal visual-spatial thinking, processing speed, long-term retrieval, and cognitive efficiency, however, were all considered weak scores because they had more than twenty points between the individual test scores. She didn't really have weaknesses in all of those areas, she had a weakness in consistently paying attention.

If one just looks at the classifications, it appears that Kaylah is an average student. One must look more closely to find possible explanations for her difficulty passing her classes in school and her symptoms of depression.

When considered with other neuropsychological tests, Kaylah was given a diagnosis of ADHD / Inattentive type. It is common for students in middle school or high school to come in to see me because of symptoms of depression or anxiety, when it is found that they have the inattentive type of ADHD. When the coping methods they used to get by in elementary school fail to work anymore, they often begin to blame themselves and believe there is something wrong with them. With treatment for ADHD, including the use of therapy to help her realize that the situation was not her fault, Kaylah improved in both her symptoms of depression and her academic performance in school.

KAYLAH	Standard Score	Classification (only for summary scores)
Verbal Ability	110	Average
Visual-Auditory Learning	88	
Spatial Relations	108	
Picture Recognition	106	
Visual-Spatial Thinking	(109)	Average
Visual Matching	81	
Decision Speed	102	
Processing Speed	(90)	Average
Short-term Memory	103	Average
Rapid Picture Naming	85	
Retrieval Fluency	112	
Long-term Retrieval	(94)	Average
Cognitive Efficiency	(98)	Average

Scores in parenthesis are considered weak

Joshua was brought in for a neuropsychological evaluation by his mother because she believed he was "a good boy," but he got in trouble a lot at school. He had been suspended for talking back to teachers, refusing to do his homework, and disrupting classes. He was being considered for special education services under the EBD category.

When tested, Joshua demonstrated significant differences between test scores for nonverbal fluid reasoning, short-term working memory, and cognitive processing speed. In addition, his summary score for cognitive efficiency was considered reliable or accurate because the two scores used to generate it were similar. Joshua's score for cognitive efficiency was significantly lower than his other summary scores. Cognitive efficiency is a combination of processing speed and short-term memory. This was a true weakness for him. The other areas represented inconsistent performance, most likely associated with inconsistent attention.

Individuals with ADHD typically have difficulty with executive functioning, which includes such things as maintaining focus, being able to prioritize tasks, and making good decisions. Executive functioning is primarily associated with the frontal lobe of the brain. Recall that earlier I called the frontal lobe our "inner adult."

Joshua was diagnosed with ADHD using these results and those of the other tests administered. He was not able to recall information (short-term memory) as well, process information as quickly, or understand nonverbal information as well as his typical peers.

The point with Joshua was that his brain worked inconsistently. He might be able to perform a task one day, but the next day, he couldn't do it. Teachers observed that he couldn't do on some days what he could do the day before, so they assumed that he wasn't trying on those days. Joshua, however, knew he was trying and just couldn't do the task. A student who is trying but accused of not trying often gives up. This is what Joshua was doing. He refused to do his homework because he believed there was no point trying. His teachers believed that he was consciously making bad decisions. For over fifty years, educational research has indicated that students will usually live up to expectations. Joshua was doing just that.

The problem with many students being served in the EBD category is that parents, teachers, and even special education teachers believe they are behaving badly on purpose. In other words, he was a "bad kid." Joshua may have given up on purpose, but the reason his brain was working inconsistently was not a conscious decision. Along with everything else, his ability to process information was extremely low, indicating that, at least in this testing application, his memory and processing speed were weaknesses for him. He missed a lot of information. Joshua needed information repeated, reinforced visually, and written down for him, as with other students with slow processing speed and/or difficulty with memory.

Joshua is doing better now, but it has not been an easy road. Medication has helped his brain work more consistently, but it has taken training to help him get used to the idea that he can complete his assignments. It's also taken awhile to convince him that he is worth our efforts.

JOSHUA	Standard Score	Classification (only for summary scores)
Number Series	71-W	
Concept Formation	105	
Fluid Reasoning (Gf)	(85)	(Low Average)
Verbal Attention	95	
Numbers Reversed	75-W	
Short-term Working Memory (Gwm)	(81)	(Low Average)

Letter-Pattern Matching	72-W	
Pair Cancellation	106-S	
Cognitive Processing Speed (Gs)	(88)	(Low Average)
Story Recall	107-S	
Visual-Auditory Learning	99	
Long-term Retrieval	103-S	Average
Visualization	100	
Picture Recognition	96	
Visual Processing	97	Average
Letter-Pattern Matching	72	
Numbers Reversed	75	
Cognitive Efficiency	68-W	Very Low

Scores in parenthesis are considered weak

Case 5

No symptoms of behavioral problems were observed, but Nicholas was having difficulty processing information and staying focused in school. He often missed hearing or remembering announcements made by his teacher.

Nicholas's scores varied significantly and his summary scores differed by fifty-five points. His strengths were nonverbal problems solving (including such things as visual puzzles and comparing shapes), long-term retrieval (learning new information), and visual processing. He was gifted in these areas.

On the other hand, his verbal ability and short-term memory were in the average range. Even though they were in the top half of the average range, they were significantly lower than his nonverbal abilities. Add to that a consideration of his average processing speed, which was in the bottom half of the average range and his cognitive fluency score in the low average range.

A lot of school personnel are not concerned when a student's lowest score is in the average range. (Cognitive fluency is not always an area of cognitive functioning that psychologists consider.) Not all cognitive batteries generate a score for it. Schools make it clear that it is not their jobs to ensure that every child meets her or his potential. It would be nice, but they don't have the resources for it. Their jobs are more related to helping students pass if they are not doing well.

Nicholas had a brain that was essentially a combination of a Cadillac and a Chevy. Here, the Chevy is an average car. If a person has a vehicle that's half Chevy and half Cadillac, one can see that it might not work efficiently. And it doesn't.

Nicholas did well at visual tasks associated with the right side of his brain, including computer design and math. When asked to write essays, however, or read long social studies assignments, he struggled to get through.

The parents and I worked with the schools to get Nicholas notes ahead of class from teachers in language-based classes such as English and Social Studies. He was also placed in an advanced math class that included a creative component in which the students worked on projects like robots. Parents learned how to access not just Nicholas' grades on the district's website, but also his daily assignments to help him stay organized.

Nicholas had previously been diagnosed with Anxiety NOS (not otherwise specified), ADHD combined type, and autism. Since Nicholas functioned fine socially and he had no stereotypical behaviors, the diagnosis of autism was not confirmed. He was not autistic but was gifted in nonverbal areas and working with a non-unified brain.

NICHOLAS	Standard Score	Classification
Processing Speed	95	Average
Cognitive Fluency	85-W	Low Average
Long-term Retrieval	128-S	Superior
Nonverbal Fluid Reasoning	140-S	Very Superior
Visual Processing	129-S	Superior
Verbal Crystalized Ability	108	Average
Short-term Memory	109	Average

Afterword

Some of the important points that I hope readers remember from reading this book include:

1) Pollyanna was right: be positive.

I couldn't really find a good place to put this into the text earlier in the book, but I think it is important for mental health.

I've had people accuse me of having my head in the sand when I promote looking on the positive side of things. I've also read books that openly criticized Pollyanna's attitude as being naïve and childish.

In the movie *Pollyanna*, the heroine is a young girl sent to her aunt in the USA when her parents die while serving as missionaries in Africa. Her parents had taught her the "glad game" while in Africa, and she brings it to her new home. In the glad game, whenever Pollyanna finds herself in a negative situation she looks for a silver lining. For example, while still in Africa, she waits for a doll to be delivered for a Christmas or birthday present. She received crutches instead. Though disappointed, she decided that she was glad she didn't need the crutches. Upon taking the "glad game" to her new home, at first her ideas are not well received because the citizens there are used to complaining about everything they don't like. Pollyanna's consistency at finding the good in life and in other people ends up changing the lives of the townspeople. Those who interact with her find a happiness they did not have before. The movie teaches that being glad or happy is a choice. Though often difficult, we can help ourselves get through life if we focus on positive attitudes rather than negative ones.

A problem with equating the negative parts of life with reality and maturity is that we get caught up the negative, even entangled with it. Our news focuses on all the bad things that happen in a day. All of the robberies, murders, cheating, and deceiving receive prime time slots on the news, giving the perpetrators attention and fame.

Concentrating on the negative aspects of our world puts stress in our lives, makes us feel victimized rather than empowered, and can make us more vulnerable to depression and anxiety. Concentrating on positive aspects of life helps relieve stress and strengthens us against depression and anxiety. We can become empowered to change ourselves and our environments.

2) Every person has strengths and weaknesses.

There is no one who has no weaknesses, and there is no one who has no strengths. Contrary to what some people would have us believe, neither money, possessions, good looks, nor winning sports events has a true impact on our worth as human beings. The worth of each individual is how well we utilize the strengths we've each been given and how we contribute to humanity.

3) Parents and teachers have impact on the physiological structure of childrens' brains.

Brains develop a blueprint of genetic material, but the fulfillment of the genetic plan is impacted by interaction with the environment. By keeping a child's environment positive, healthy, and enriched with opportunities for learning, we give children the best possible chance for success. The opportunities don't need to be costly. A public library can be a fantastic place to open up the world for your child. The internet is also wonderful, but do not turn your child loose on the internet. There are predators there.

4) The vast majority of children will behave well if they can.

Public schools and parents struggle with the behaviors of children, whether it's not doing homework, disrupting class or the home, or hurting others. Children who have come to me for an evaluation have proven to me that there is usually an underlying reason for the behaviors. They are frustrated, angry, or have little confidence in their own worth as a human being.

Sometimes the cause is as easy as them being hungry. Sometimes the reasons can only be discovered by an assessment. But we need to recognize that these children are part of our amazing world. It is our responsibility to help discover what may be causing the behaviors so that hopefully they can be changed. Most importantly, these approaches need to be put in place early, before bad behaviors become part of the child's brain structure.

5) Our mothers were also right.

The easiest way to foster the growth of your childrens' brains is to follow what mothers have been telling children for generations: eat right (plenty of protein and vegetables), get plenty of sleep, and be kind.

Acknowledgements

These projects cannot come to fruition without assistance from many different directions. I want to thank Gary Lindberg for giving me the confidence to go forward. He knows all of the aspects of publishing, and without a collaborative effect, nothing happens.

I also want to thank Dr. Betty Gridley and Dr. Ray Dean at Ball State University, and the university in general. They pushed when necessary and channeled me to focus when my attention was on too many areas at once. Dr. Gridley had to say K.I.S.S. (Keep It Simple Stupid) more than once. The neuropsychology program at Ball State is amazing, and my experience there started me down the road to a career I enjoy.